The Camera Assistant's Manual

SECOND EDITION

The Camera Assistant's Manual

SECOND EDITION

David E. Elkins

Butterworth—Heinemann

Boston Oxford Johannesburg Melbourne
New Delhi Singapore

Focal Press is an imprint of Butterworth-Heinemann.

Recognizing the importance of preserving what has been written, Butterworth-Heinemann prints its books on acid-free paper whenever possible.

Library of Congress Cataloging-in-Publication Data

Elkins, David E.
The camera assistant's manual / David E. Elkins.—
2nd ed. p. cm.
Includes bibliographical references and index.
ISBN 0-240-80242-X (pbk.)
1. Cinematography—Handbooks, manuals, etc. I. Title.
TR850.E37 1996
778.5´3—dc20 96-5429
CIP

British Library Cataloguing-in-Publication Data
A catalogue record for this book is available from the British Library.

The publisher offers discounts on bulk orders of this book.
For information, please write:
Manager of Special Sales
Butterworth-Heinemann
313 Washington Street
Newton, MA 02158-1626
Tel: 617-928-2500
Fax: 617-933-2620

For information on all Focal Press publications available, contact our World Wide Web home page at:
http://www.bh.com/bh/

10 9 8 7 6 5 4 3 2

Printed in the United States of America

To my father,
who always believed in me

Table of Contents

Preface

As a motion picture camera assistant, you must be constantly aware of many things happening around you during the performance of your job. There are many responsibilities and duties that a camera assistant should know about. This book is intended to be a guide for the individual who would like to learn to become a camera assistant. When I first started in the film industry, there was no book that explained how to do the job of a camera assistant. Even while I was in film school, there was no course dealing with this specific area of production. All my training came from on-the-job work experience. I hope that with this book any student or beginning filmmaker who wishes to become a camera assistant will find it a little easier to learn the job duties and responsibilities. With the knowledge obtained from this book, it should be easier to obtain your first job because you will at least know the basics and should have no trouble applying them to actual shooting situations.

In preparing the material for this book, I began with a description of the basics of cinematography. Many readers of this book may have no previous photography or cinematography experience. This basic introduction will help beginners to understand much of the terminology used throughout the book. Following the basics of cinematography, there is a description of the chain of command within the camera department and how each member works with and relates to the others. Following this I chose to cover the job responsibilities of a Second Camera Assistant, and then move on to the First Camera Assistant. My reason for this is that when most people first start in the camera department, they start as a Second Camera Assistant. Once they have worked at that position for some time, they move up to First Camera Assistant. The length of time spent at each position depends on each person's situation or preference. Following the chapter on the First

Camera Assistant is a chapter that discusses problems that may arise, and what you should do to either correct or prevent them. This is an important part of the job of a camera assistant. Chapter 6 contains illustrations of most of the currently used cameras and magazines, and Chapter 7 contains some tips and guidelines on what to do before you have the job, once you are working, and after the job is over. All of the information in these chapters is based on my experience as a camera assistant.

The appendices cover five areas: film stock, equipment, checklists, tools and accessories, and tables. Appendix A is a complete listing of all film stocks available from the various manufacturers at the time of publication. It lists the recommended exposure index (E.I.) ratings for each stock for different lighting conditions. Appendix B lists the name of the most common pieces of equipment that you will be working with and should know about. Appendix C contains checklists for camera rental items, filters, and expendables that are usually needed on each production. Appendix D lists the basic tools and accessories that a camera assistant needs to do the job. Appendix E contains many useful tables that you may need to refer to in the course of your job.

Following the appendices is a list of recommended books for the camera assistant who would like to learn more about the film industry. The glossary lists many of the key terms used in the book and their meanings. The items on the expendable list, camera rental items, and various filters are included in the glossary.

Acknowledgments

In preparing this book I have used the knowledge and information received from many friends and colleagues. I would like to extend my deepest thanks to all who have contributed their ideas and helped me in the preparation of this book and throughout my career. Special thanks go to many people who have been especially helpful. To the many Directors of Photography with whom I have worked, including Eric Anderson, Ken Arlidge, Arledge Armenaki, Richard Clabaugh, Steve Confer, Robert Ebinger, Paul Elliott, Tom Fraser, Victor Goss, Kim Haun, Gerry Lively, Paul McIlvaine, Tony Palmieri, Marvin Rush, Aaron Schneider, Stephen Sealy, Randy Sellars, John Schwartzman, Tom Spalding, Joe Urbanczyk, Richard Walden, and Mark Woods, thanks for all your help and understanding. To my fellow camera assistants who have shared many of their ideas, Chip Bailey, Bobby Broome, Michael Cardone, Gina DeGirolamo, Buddy Fries, Chris Goss, Mike Hanly, Beth Horton, David Huey, Duane Manwiller, Michael Martino, Kelly McGowan, Michael Millikan, Steve Monroe, Chris Mosely, Haywood Nelson, Chris Poncin, Maricella Ramirez, Brian Sorenson, Scott Ressler, Tim Roarke, Pat Swovelin, Ray Wilbar, and all the others whose names I cannot remember, thanks for the many enjoyable hours of working together. To all the other crew members on the various productions on which I have worked, thanks for making each workday a little more interesting and enjoyable.

To the employees at the various camera rental houses, Birns & Sawyer, Clairmont Camera, Keslow Camera, Otto Nemenz, Int., Panavision, Inc., Panavision Hollywood, and Ultravision, thanks for all the help before and during the many productions we have done together. To Brian Lataille, thanks for the information about the Steadicam. To Larry Barton of Cinematography Electronics, thanks for providing me with much needed information. To Bill Russell at Arriflex; Frank Kay, Dan Hammond, and Jim Roudebush at Panavision; Matt Leonetti at Leonetti Camera; Philip Kiel at Photosonics; Gary Woods at Aaton; and Grant Loucks at Alan Gordon Enterprises, thanks for letting me use the various illustrations included in the book. To Rudy at the Paramount Studios Camera Department, thanks for your help. Thanks to Jeremy Boon for drawing some of the illustrations that I could not find anywhere else. To Nicole Conn for allowing me to use the names "Demi Monde Productions" and "Claire of the Moon" in the various examples and illustrations in the book, thanks for letting me have so much fun on your film. Thanks to Pat Swovelin for his 2nd A. C. Pointers.

To the entire staff of Columbia College-Hollywood, especially Allan Rossman and Dianne McDonald, thank you for your help and guidance. To Mike Hanly for giving me some insight on dealing with a camera rental house, thanks for your help. To Richard Clabaugh, who spent many hours reading the final manuscript of the first edition and offering his suggestions, thanks for your time and effort. To Karen Speerstra, Sharon Falter, Marie Lee, Valerie Cimino, and Tammy Harvey at Focal Press, thanks for making the writing of this book a little easier because of your help and guidance.

Finally, and most important, thanks to my family and friends for being understanding and supportive. To Peter, Janet, Fred, Doreen, Paul, Karen, and Candy, thanks for always being there. To Jan, thanks for your love and understanding.

Introduction

The process of motion picture photography started when George Eastman introduced the first 35mm film in 1889, and Thomas Edison and his assistant W.K.L. Dickson designed the Kinetograph and Kinetoscope around 1889. The Kinetograph was used to photograph motion pictures and the Kinetoscope was used to view them. These early pieces of equipment were very basic in their design and use. As film cameras became more complex, a need developed for specially trained individuals to work with this new technology and equipment. Two of these individuals became known as the First Camera Assistant (1st A.C.) and the Second Camera Assistant (2nd A.C.).

One of the most well known of the early cinematographers was Billy Bitzer, who shot most of the films of director D.W. Griffith. As a cameraman he did all of the jobs himself, carrying the equipment, setting it up, loading film, and so on. In 1914 D.W. Griffith hired an assistant to work with the cameraman. This assistant was called a camera boy, and his job was only to carry all of the equipment for the cameraman. Each morning, the camera boy would move all of the equipment from the camera room to wherever the scenes were being shot for the day. There was a lot of equipment, and many trips back and forth were required to get everything in place. In addition, the camera boy was required to take notes of what was being shot. There were no script supervisors at that time.

Around 1916, cameraman Edwin S. Porter asked for an assistant, after returning from a long location shoot. This camera assistant had some additional duties that the camera boy did not have. Because all of the early cameras were hand-cranked, the assistant had to count the number of turns of the crank and keep a log of the number of frames shot. Other duties included slating the scene, keeping track of footage, loading and unloading film, carrying and setting up the equipment, and anything else that he may have been asked to do.

Because of these two early cameramen having an assistant, a new position was created within the camera department. I am guessing that many of the techniques of these early cameramen and assistants were passed on to other cinematographers, and they developed into the very specific job duties performed today by the 1st A.C. and 2nd A.C. Some of those early job duties are still part of the camera assistant's job requirements. Because this was such a new technology, the early camera assistants had no one to learn from so they probably set most of the guidelines for performing their specific jobs. Each had specific responsibilities but was also capable of doing the other's job if necessary.

Today, a beginning filmmaker has a wide choice of places to get the training to work as a camera assistant. There are many colleges and universities that offer a complete curriculum dealing with motion picture production. In addition to the larger institutions, there are many smaller colleges and trade and technical schools that offer film classes. The beginning filmmaker may know someone in the film industry who is willing to train him or her, and give that important first break. I know many film professionals who have never attended film school but obtained their training and experience by starting out working on various productions. There is no right or wrong way to gain the experience. It is a matter of which way is best for you.

If you choose to attend some type of film school, the best way to gain actual production experience is to work on as many student film productions as possible. Even though these productions are done on a much smaller scale than most professional productions, the basics will be the same and you can apply what you have learned in your film classes. When you start looking for that first professional job, any experience, even if it is on a student production, increases your chance of getting a job. For those who do not wish to go to film school, or perhaps cannot afford the cost, it may be a little more difficult to obtain that first job. If you have an acquaintance or relative in the film industry, it may be a little easier. For me, film school was a valuable and rewarding experience. I was hired on my first production as a production assistant only one month after completing film school. That position led to my first job as a 2nd A.C. on the same film. The film crew was going to do some second unit shooting and needed a 2nd A.C. to load magazines and keep camera reports. The production manager had been a classmate of mine in film school, and he recommended to the Director of Photography (D.P.) that I be given the chance to work as the 2nd A.C. on the second unit. The D.P. gave me the opportunity to prove that I could do the job, and this led to my first job on a feature film as 2nd A.C.

You must be willing to work hard, not only at getting the job but also once you have the job, to prove that you are capable of handling it. If you have been in film school recently, an excellent way to learn about available jobs is to talk to your instructors. Ask them if they know about any productions that you may be able to work on. You also should stay in contact with other film students who were in your classes. There are also a few publications that come out daily or weekly that deal strictly with the film industry. These publications usually have a listing of productions being done now or sometime in the future that often contains phone numbers or addresses to obtain more information about each production. The three publications that carry the most information are *Daily Variety*, *Drama Logue*, and *The Hollywood Reporter*. They are available at newsstands in most major cities and also by subscription.

When you first try to get a job on a film, you may be asked to work for little or no money. The production company may be just starting out and have only enough money for the basic expenses. If you can afford to take such a job, it is an excellent way to get some experience. Three of my first jobs as a camera assistant were without pay, but they helped me to get paying jobs later because I had proved that I could do the job and was not afraid to work long, hard hours. Not everyone will find it necessary to work for free. I mention it only so that you know what you might find when you first start looking for work. The important thing to remember is not to get discouraged and give up. The film industry is a very competitive business; breaking into it may take awhile. If you don't get the first few jobs you apply for, keep trying. If you want a job bad enough and are willing to work, you will eventually find one.

Once you do start working in the industry, always stay in contact with people with whom you have worked in the past. Call them periodically just to say hello and find out what they are doing. They may be working on a production that needs additional crew members. Also, if you are working on a production that needs additional people, be sure to let other film professionals know about it. This process of keeping in touch with other film crew people is called *networking* and is probably one of the best ways to get jobs. Many of my jobs came from recommendations from people with whom I have worked on other productions. Also, many D.P.s will call me back to work with them on other productions.

Good luck to all the aspiring camera assistants who read this book. I hope that you find the motion picture industry to be as exciting and rewarding as I have. And don't forget, work hard but have fun too.

1

Basics of Cinematography

The motion picture industry uses many terms and principles that are not used anywhere else. In order to perform your duties as a camera assistant, you need to be aware of these terms and the basics of cinematography. You will hear many of these terms and principles in the day-to-day performance of your job. I am presenting them here and briefly explaining them so that you will at least have a basic understanding of them. By introducing you to them in this chapter, I hope that it will make it easier for you as you read the book, and also make it easier for you the first time you step onto a film set. To my knowledge all of this information is true and accurate. If you would like a more in-depth discussion about any of this information, you may consult any of the books listed in the recommended reading at the end of this book.

FILM FORMATS

The term *format* may be used to indicate a few different things in the motion picture industry. When we say film format, we are referring to the size of the film stock being used for shooting. The two primary film formats used for shooting all filmed productions, and the formats that you will most often be working with as a camera assistant, are 16mm and 35mm. Almost all professional cinematography is shot using either 16mm film or 35mm film. The 65mm/70mm format is a popular release print format but is not used much for production.

In order for the film to be moved through the camera, it contains perforations. The perforations may also be referred to as *perfs* or *sprocket holes.* These are equally spaced holes that are punched into the edges of the film so that it can be transported through the camera at a constant speed. Sixteen millimeter film contains two perforations per frame on each side of the film. There are 40 frames per foot of 16mm film. The standard 35mm format contains four perforations per frame on each side of the film, and there are 16 frames per foot (Figures 1.1 and 1.2).

The 65/70mm format is a popular release print format. Many films that are photographed on 35mm film are enlarged to 65/70mm for release to the theaters. A larger negative will result in a sharper, clearer picture when projected on the big screen. Figure 1.3 shows the 65/70mm film frame.

SYNC SPEED

When we use the term *sync speed*, we are referring to the speed at which the film moves through the camera in order to create the illusion of normal motion when viewed. In the United States normal sync speed is 24 frames per second (24 f.p.s.). This means that 24 frames of film travels through the camera each second during filming and pro-

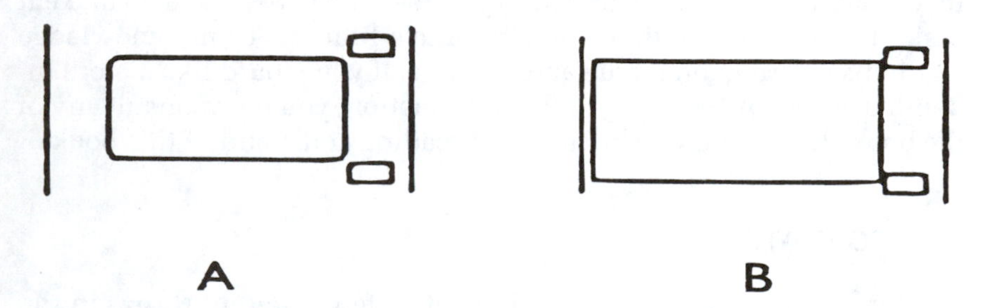

Figure 1.1 (A) Regular 16mm film frame. (B) Super 16mm film frame.

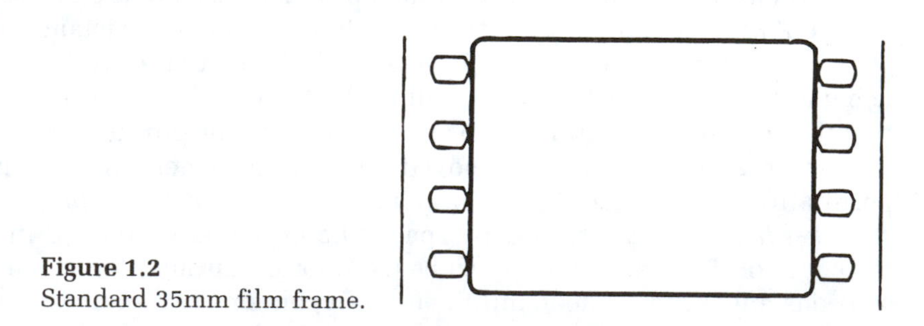

Figure 1.2
Standard 35mm film frame.

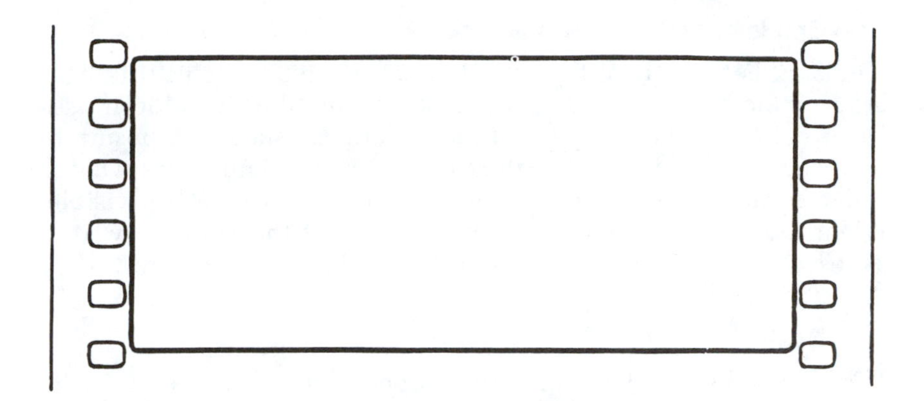

Figure 1.3 The 65mm full frame.

jection. In Britain and Europe, sync speed is 25 f.p.s. Anything filmed at a frame rate less than sync speed will have the illusion of high speed when projected. Anything filmed at a frame rate more than sync speed will have the illusion of slow motion when projected. For 16mm cinematography, at sync speed the film will travel through the camera at the rate of 36 feet per minute. For standard 35mm cinematography, at sync speed the film will travel through the camera at the rate of 90 feet per minute. For 3-perf, 35mm cinematography, at sync speed the film will travel through the camera at the rate of 67.5 feet per minute.

SYNC AND MOS

The two types of motion picture filming are sync and MOS. During filming when we are recording synchronous sound, such as dialogue, along with the picture, this is referred to as *sync* filming. When we are filming without recording synchronous sound, this is referred to as *MOS* filming.

FILM STOCK

Any piece of motion picture film stock is made up of three main components.

Emulsion

Emulsion is the part of the film that is sensitive to light. It is light brown (color film) or light gray (black & white film) in color. It uses silver halide crystals suspended in a gelatin substance. Exposure to light causes a change in the silver halide crystals and forms what is called a *latent image*, which means an image that is not yet visible. When the film is developed and processed at the laboratory, it is exposed to various chemicals, forming a visible image (Figure 1.4).

Base

The base is the flexible, transparent support for the emulsion. In the early days of filmmaking it was made up of highly flammable cellulose nitrate, but today it is cellulose acetate, which is not flammable and is much more durable and long lasting. The base does not play a part in forming the image on the film, but acts only as a support for the emulsion (see Figure 1.4).

Anti-halation Backing

Anti-halation backing is the dark coating painted on the back of the base. It is there to prevent light from passing through the film, bouncing off the back of the camera, and causing a flare or flash in the image (see Figure 1.4).

TYPES OF FILM

The two main types of film that are available for shooting are negative and reversal.

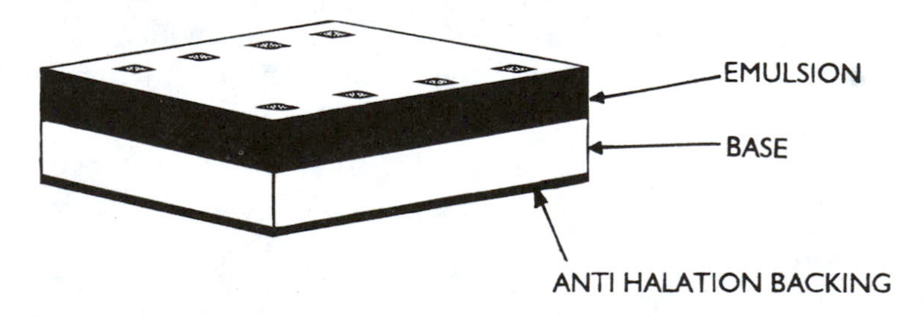

Figure 1.4 Enlarged cross section of a piece of film (not drawn to scale).

Negative film produces a negative image when developed, in which blacks are white, whites are black, and each color is its opposite or complementary color. A positive print must be made from the negative so that you have something that is suitable for projection and viewing. Recently it has become very common to directly transfer the negative to videotape for editing and projecting. During the transfer process, the colors are switched to their positive image electronically. One of the primary advantages of using negative film is the ability to make any exposure corrections during the laboratory printing process. Negative film is also better suited to making a large number of copies, which is done for feature films that are shown in many different theaters at the same time. For all professional cinematography, negative film is the most commonly used.

Reversal film produces a positive image when developed, and you are able to project the camera original without making a print. A good example of reversal film is Super 8mm home movie film. Most slide film is also reversal-type film. It is possible to make a print from reversal, but it is not as well suited to making multiple copies as negative film is.

FILM STOCK MANUFACTURERS

The two companies that manufacture film stock for professional motion picture productions are Eastman Kodak and Fuji. Since the most widely used film in the motion picture industry is manufactured by Eastman Kodak, most of the examples in this book will use Eastman Kodak Color Negative Film. Eastman Kodak uses a series of numbers to designate each specific film stock, and they are commonly written in the following manner: 5296-197-1102 or 7293-032-1902. The numbers 5296 and 7293 refer to the kind of film it is, 197 and 032 are the emulsion numbers, and 1102 and 1902 are the roll numbers of that particular emulsion. Usually when we speak of the film's emulsion number, we include both the emulsion number and the roll number. To distinguish between 16mm and 35mm film stock, Eastman Kodak designates any film stock that begins with the number 72 as 16mm and any film stock that begins with the number 52 as 35mm. So, for the example given, 7293-032-1902 is 16mm film stock and 5296-197-1102 is 35mm film stock.

Fuji designates its film stocks using the prefix 85 for 35mm film stock and 86 for 16mm film stock. For example, 8561-987-654 is 35mm film and 8631-298-157 is 16mm film. See Appendix A for a complete listing of all currently available motion picture film stocks.

STOCK PACKAGING SIZES

…tion to knowing which film stock to use for your production, …st know what sizes are available for the rolls of film. The size …e roll will be based upon what camera system you are using. Certain cameras only accept specific size rolls of film. Sixteen millimeter film is available on rolls ranging from 100 feet in length to 1200 feet in length. Thirty-five millimeter film is available on rolls ranging from 100 feet to 2000 feet. See Appendix A for a complete listing of all currently available motion picture film stock packaging sizes.

Film stock may be packaged on a daylight spool or on a plastic core. Daylight spools allow you to load or unload the film in daylight or subdued light, while film wound onto a plastic core must be loaded or unloaded in complete darkness. See Figure 1.5 for an illustration of 16mm and 35mm daylight spools. See Figure 1.6 for an illustration of 16mm and 35mm plastic film cores.

COLOR TEMPERATURE AND COLOR BALANCE

For professional cinematography, each light source is considered to be a different color and therefore has what is referred to as a different *color temperature*. Scientists take an ideal substance, which they refer to as a "black body" and heat it, and then measure its temperature as it emits different colors of light. This temperature is called *color tem-*

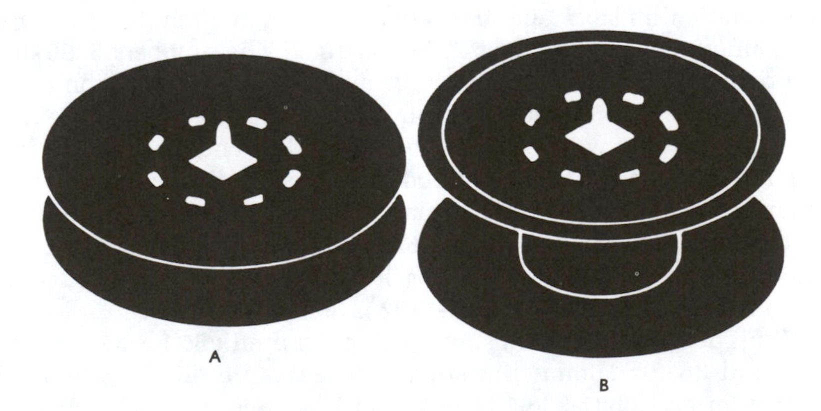

Figure 1.5 (A) 16mm daylight spool. (B) 35mm daylight spool.

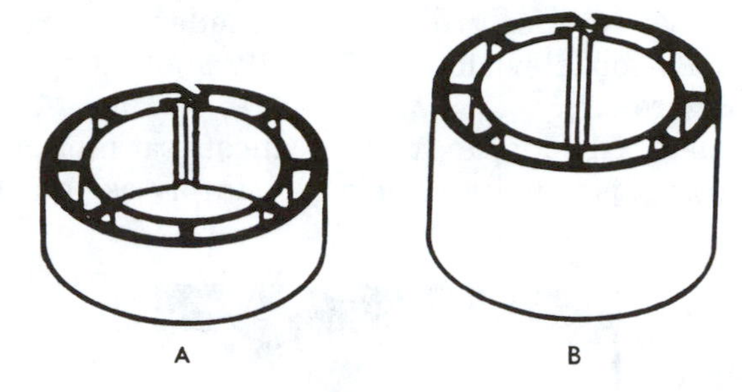

Figure 1.6 (A) 16mm 2″ plastic core. (B) 35mm 2″ plastic core.

perature. Color temperature is measured in degrees Kelvin (K), which is a temperature scale used in physics.

The two main types of light sources for professional cinematography are daylight and tungsten light. Daylight has a color temperature of approximately 5600 degrees Kelvin, written as 5600°K. Tungsten light has a color temperature of approximately 3200 degrees Kelvin, written as 3200°K. When we refer to any particular film stock, we say that it is either daylight balanced or tungsten balanced.

Daylight balanced film can be shot in daylight without making any adjustments to the light or adding any filters to the camera to correct the color temperature. Tungsten balanced film can be shot in tungsten light without making any adjustments to the light or adding any special filters to the camera to correct the color temperature. You may use either film in the opposite type of light, but you must place a specific filter on the camera in order to correct for the difference in color temperature. These specific filters will be discussed in the section on filters later in this chapter.

ASPECT RATIOS

The shape of the frame is expressed as a ratio of its width to its height. This is referred to as the *aspect ratio* of the image. The three most commonly used aspect ratios for filmed productions are 1.33:1, read as "one three three to one," 1.85:1, read as "one eight five to one," and 2.40:1, read as "two four oh to one." The 1.33:1 aspect ratio may also be referred to as *academy aperture*. It is 1.33 times as wide as it is high. Many of the early motion pictures were shot using this aspect ratio. Present day television still uses the academy aperture, and any films shot strictly for television are usually shot using the academy aspect ratio (Figure 1.7).

The standard aspect ratio for most theatrical motion pictures is 1.85:1. This format is usually referred to simply as "one eight five."

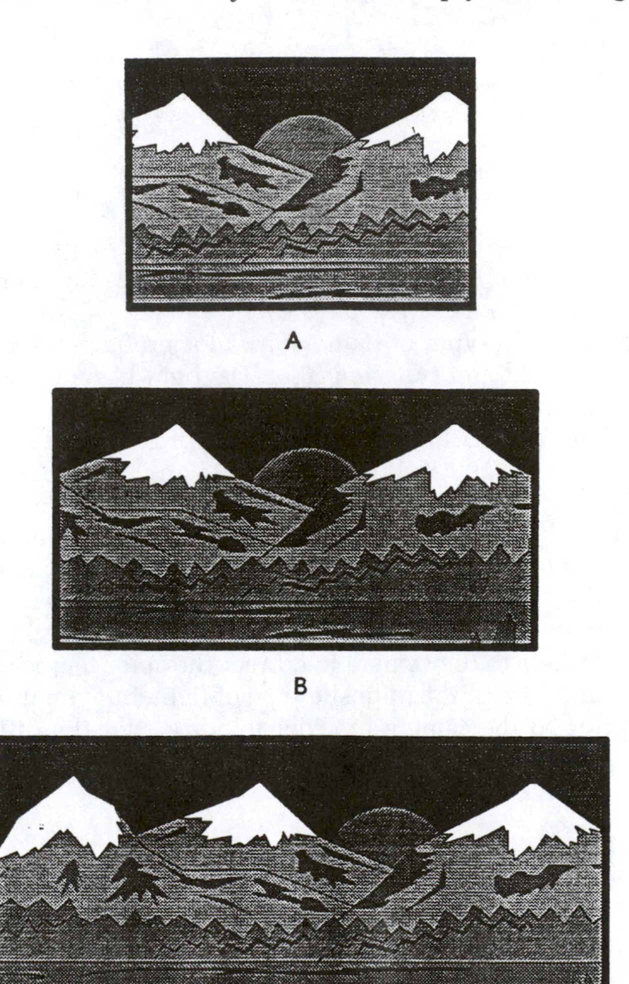

Figure 1.7 Comparison of 1.33, 1.85, and 2.40 aspect ratios: (A) 1.33:1, (B) 1.85:1, (C) 2.40:1. (Courtesy of Panavision, Inc.)

This wider format is obtained by chopping off the top and bottom portions of the academy aperture to give an image that is exactly 1.85 times as wide as it is high (see Figure 1.7).

The 2.40:1 aspect ratio is also called *anamorphic*, and the image is 2.40 times as wide as it is high. To obtain this aspect ratio, a special anamorphic lens is used that squeezes the wider image onto a standard 35mm frame of film. When the image is projected, it is unsqueezed to produce the wide screen image (see Figure 1.7).

In addition to the above-named aspect ratios, there are others that are available. One of these is 1.66:1, which is the common aspect ratio for European released films. Another aspect ratio is 1.78:1, which was developed for the new high definition televisions (HDTV). These new televisions have a screen that is almost the same aspect ratio as the standard 1.85:1 movie screen (Figure 1.8).

Panavision has developed a camera system that is designed to reduce film waste. Instead of the standard four perforations per frame, this system uses three perforations per frame of 35mm film, which results in a 25% savings in film use. When shooting 1.85:1 aspect ratio

A

B

Figure 1.8
(A) 1.66:1 aspect ratio.
(B) 1.78:1 aspect ratio (HDTV).
(Courtesy of Panavision, Inc.).

for theatrical films, and using 4-perf 35mm film cameras, the top and bottom part of the film frame is wasted. By using 3-perf cameras there is almost no wasted film (Figure 1.9).

EXPOSURE TIME

The length of time that each frame of film is exposed to light is called the *exposure time.* At sync speed, film is traveling through the camera at the speed of 24 frames per second. This means that each frame is being exposed for 1/24th of a second. For one-half the time, the film is being moved in and out of position in an area of the camera known as the *gate*, and for one-half the time it is being exposed to light. One half of 1/24th of a second is equal to 1/48th of a second. For conve-

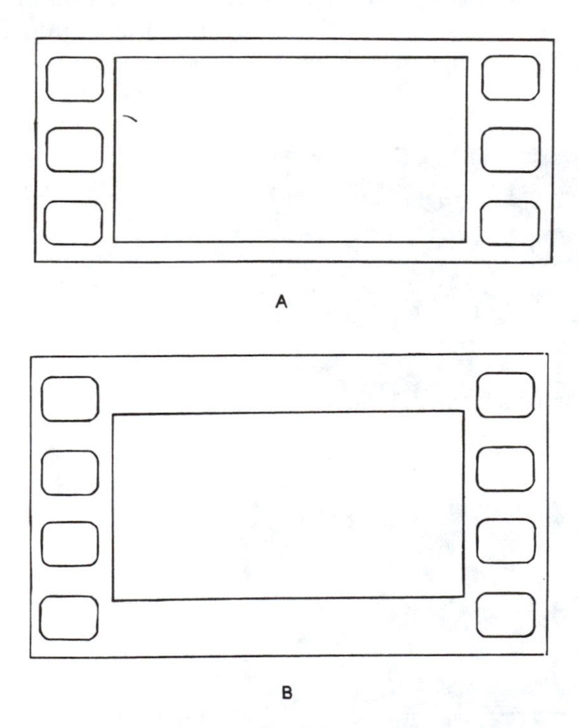

Figure 1.9 (A) 1.85 film frame when using 3-perf cameras. (B) 1.85 film frame when using 4-perf cameras. (Courtesy of Panavision, Inc.)

nience this is rounded to the nearest tenth, making it 1/50th of a second. This is the actual amount of time that each frame of film is being exposed to the light. Therefore, at sync speed we say the standard exposure time for all motion picture photography is 1/50th of a second.

F-STOPS AND T-STOPS

All motion picture lenses contain some type of adjustable iris or diaphragm to control the amount of light that enters the lens and strikes the film. A wide opening allows more light to strike the film than a small or narrow opening. The measurement of the width of this opening is called an *f-stop*. The standard series of f-stop numbers is

1, 1.4, 2, 2.8, 4, 5.6, 8, 11, 16, 22, 32, 45, ...

It is important to remember that the f-stop numbers go infinitely in both directions.

All lenses are marked along the barrel of the lens with these f-stop numbers. By turning the diaphragm or iris adjustment ring on the lens barrel to a specific number, you are adjusting the size of the iris diaphragm within the lens and controlling how much light gets through to the film. Each f-stop admits half as much light as the f-stop before it. It is also important to remember that as the f-stop numbers get larger, the opening of the iris diaphragm gets smaller.

In addition to f-stops, professional cinematography uses another form, which is called a *t-stop*. An f-stop is a mathematical calculation based upon the opening of the diaphragm opening. A t-stop is a measurement of the actual amount of light transmitted through the lens at a particular diaphragm opening. F-stops and t-stops are discussed further in Chapter 4, First Camera Assistant.

EXPOSURE INDEX OR ASA

All motion picture film is sensitive to light in varying intensities. The measurement of how sensitive a particular film stock is to light is called its *exposure index (E.I.)* or *ASA number*. The larger the E.I. or ASA number, the more sensitive the film is to light, and the less light it needs for a proper exposure. The smaller the E.I. or ASA number, the less sensitive the film is to light and the more light it needs for a proper exposure. Similar to the series of f-stop numbers, there is a series of E.I. or ASA numbers used to rate the film's light sensitivity. These numbers are

12, 16, 20, 25, 32, 40, 50, 64, 80, 100, 125, 160, 200, 250, 320, 400, 500, 650, 800, 1000...

By looking closely at this list of numbers you will see that the values double every three numbers. This means that every three numbers equals a change of one full f-stop, or each ASA value is equal to one third of a stop.

As was mentioned earlier, in the discussion of f-stops, a change of one full f-stop either doubles the amount of light or cuts it in half. So, by doubling or dividing in half the ASA number, it is the same as doubling or dividing in half the amount of light. As an example, the same amount of light that gives you an exposure of f/4 at ASA 200 will require an f/5.6 at ASA 400 or an f/2.8 at ASA 100.

The exposure index (E.I.) or ASA is determined by the film's manufacturer based on extensive testing of the film. This number is what the manufacturer feels will give the "best" exposure of the film. Each film can label will show the recommended E.I. or ASA rating for the film stock, for both daylight and tungsten light. The ultimate decision on what E.I. or ASA to rate the film is up to the Director of Photography, and is usually based on his or her experience in using the particular film stock.

EXPOSURE METERS

In order to determine the correct f-stop or t-stop to set on the lens for shooting, we measure the intensity of the light with an *exposure meter* or *light meter*. The two basic types of light meters used for measuring the exposure of an object are incident meters and reflected meters.

Any light that is falling on an object is called *incident light* and is measured with an *incident light meter*. The meter contains a white, translucent dome called a *photosphere*, which is mounted over a light sensor. The photosphere simulates a three-dimensional object, such as a human face, and averages the light falling on an object from all angles. The recommended and accepted way to use an incident light meter is to stand at the position of the subject being photographed and point the photosphere toward the camera.

Any light that bounces off or is reflected by an object, is called *reflected light* and is measured with a *reflected meter*. The light that is reflected by an object is based on the color and texture of the object. A white object will reflect more light than a black object. A smooth object will reflect more light than a textured object of the same color. The area in which a reflected meter actually reads the light is called the *angle of acceptance*. The most commonly used reflected light meters are called *spot meters* and have a very narrow angle of acceptance, usually around 1°. The recommended and accepted way to use a spot meter is to stand at the position of the camera and point the meter toward the subject being photographed (Figures 1.10–1.13).

THE CAMERA

All motion picture cameras are made up of many different components and accessories. Each camera manufacturer has its own specific design for the various parts, and these parts are usually not interchangeable from one make of camera to another. A basic camera may be made up of the following components: gate, shutter, inching knob, viewing system, motor, lens, and magazine. There are many more specific components that are used on all motion picture cameras that you will learn as you work as a camera assistant. For now I will discuss only these basic parts.

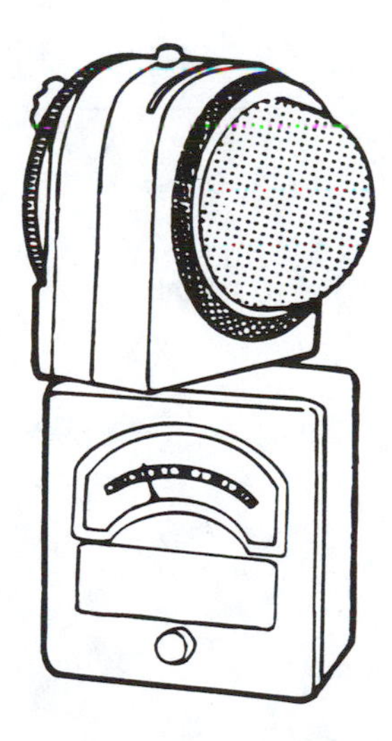

Figure 1.10
Spectra incident light meter.

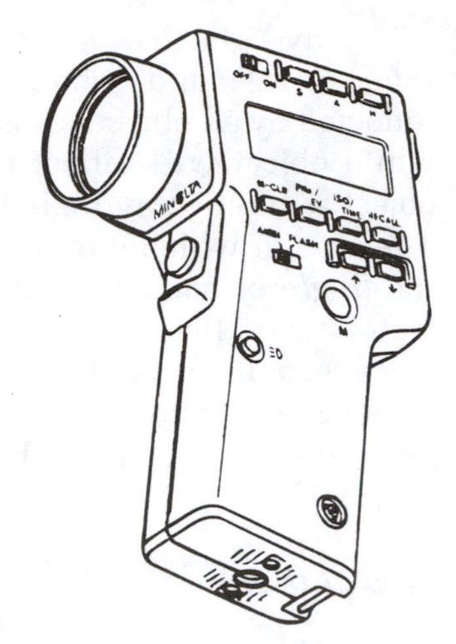

Figure 1.11
Minolta reflected (spot) meter.

Gate

The gate is the opening in the camera, which allows light passing through the lens to strike the film. It may also be referred to as the *aperture*. We sometimes refer to the entire area within the camera

Figure 1.12
Using the incident light meter.

Figure 1.13 Using the reflected (spot) light meter.

where the film is exposed as the gate. As the film moves through the gate, it moves by a process known as intermittent movement.

Intermittent Movement

To the human eye, it appears that the film is constantly moving as it travels through the camera. Actually, as the film moves through the camera, each frame is held in place in the gate for a fraction of a second before it moves on and is replaced by another frame. While the film is held in the gate for this fraction of a second, it is being exposed to light. The process of holding one frame of film in the gate, and then moving it, so the next frame is brought into position, is called *intermittent movement.* This process of starting and stopping the film happens at the rate of 24 frames per second, which we learned earlier is called sync speed.

During the threading of the film in the magazine or in the camera, you usually set a portion of the film to a specific length between the gears or teeth of the magazine or camera movement. This length of film is called the *loop,* and it acts as a buffer between the intermittent movement and the entire roll of film (see Figure 1.14).

There are four components to the gate area that work together to make this intermittent movement happen.

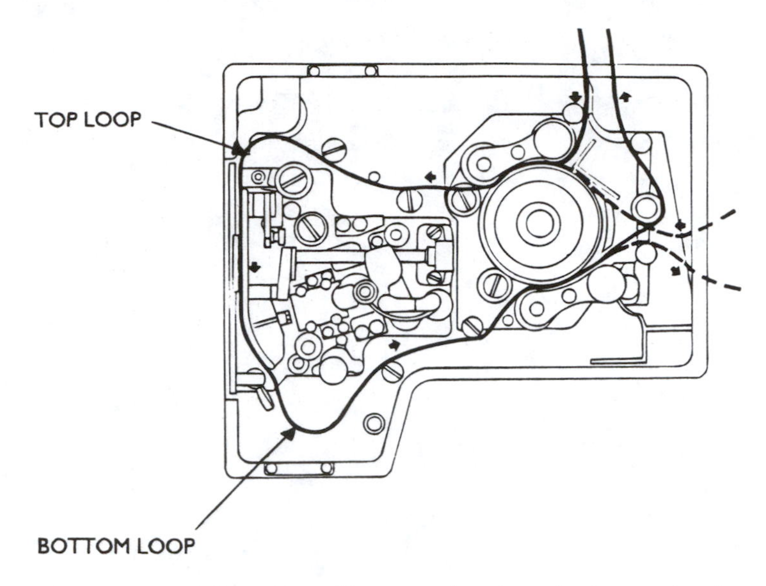

Figure 1.14 Threading diagram showing the loops in the Panavision camera. (Courtesy of Panavision, Inc.)

Pull Down Claw: In order to move the film, a small hook engages into a perforation in the film and pulls it through the gate. This small hook is called the *pull down claw*. Each camera contains some type of pull down claw to move the film (see Figure 1.15).

Registration Pin: Once the pull down claw pulls the film into the gate so that it may be exposed, it must be held perfectly still during this exposure process. A small metal pin engages into the film's perforation and holds it in place so that it may be exposed. This small pin is called the *registration pin* (see Figure 1.15).

Aperture Plate: The metal plate that contains the opening or gate through which light passes to the film is called the *aperture plate*. The opening may be called the *gate* or the *aperture* (not the same as the lens aperture or f-stop), and is usually the same size as the aspect ratio being used. The term *aperture* means "an opening," and we usually speak separately of lens apertures and camera apertures (see Figure 1.15).

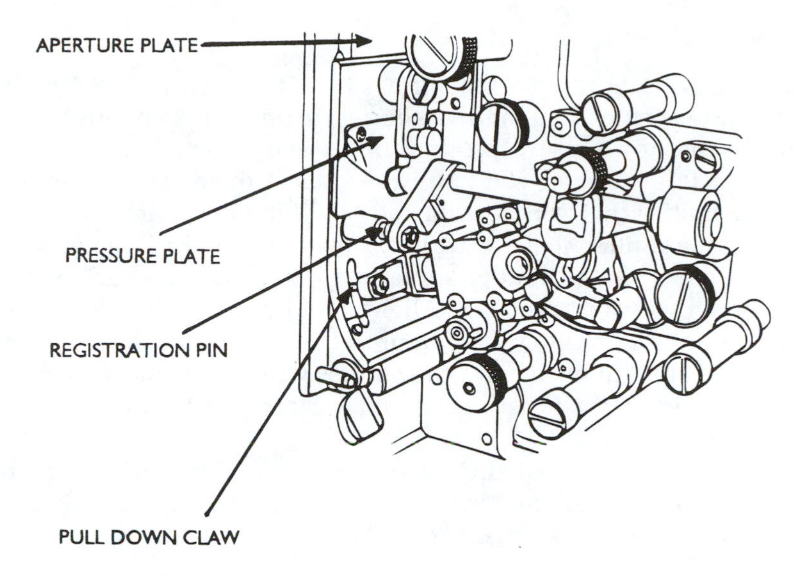

Figure 1.15 Panavision gate area showing various components. (Courtesy of Panavision, Inc.)

Pressure Plate: The area where the film is being held in the gate during exposure is called the *film plane* or *focal plane.* In order to keep the film flat against the aperture plate during exposure, there is a small metal plate located behind the film, which pushes it against the aperture plate and keeps it flat and steady in the film plane. This small piece of metal is called the *pressure plate,* because it puts pressure against the film (see Figure 1.15). When referring to the gate, we usually mean that area in the camera that contains the pull down claw, registration pin, aperture plate, and pressure plate.

Shutter

A basic definition of the shutter is that it is an on/off switch for the light striking the film. The shutter is mechanically linked to the other parts of the intermittent movement so that its timing is synchronized with the movement of the pull down claw and registration pin. As the pull down claw moves the film into position, the shutter is in the closed position so that no light strikes the film. Once the frame of film is in place and being held by the registration pin, the shutter is then in the open position so that the light may strike the film and create an exposure.

Shutter Angle

The opening in the shutter that allows the light to strike the film and create an exposure is called the *shutter angle*. The standard shutter angle for motion picture production is 180°. On all professional motion picture cameras, you will have either a fixed 180° shutter or a variable shutter that can be adjusted to different shutter angles. By changing the shutter angle, you affect how long the film is exposed to light. By reducing the shutter angle, you reduce the amount of time that the film is being exposed to light, and by increasing the shutter angle, you are increasing the amount of time that the film is being exposed to light. In most cases the shutter will be one of two types, a standard solid 180° shutter or a double-bladed 180° shutter (Figures 1.16 and 1.17).

In addition to the rotating mirror shutter, all Panavision cameras contain a focal plane shutter, which controls the light striking the film, while the mirror shutter is only for the reflex viewing system.

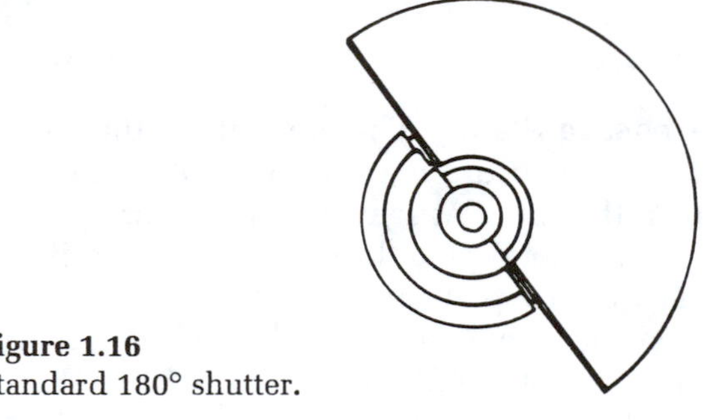

Figure 1.16
Standard 180° shutter.

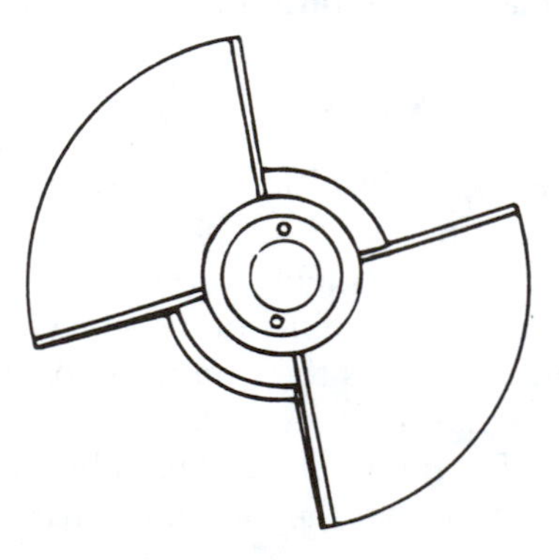

Figure 1.17
Double-bladed 180° shutter.
(Courtesy of Panavision, Inc.)

Inching Knob

Most professional motion picture cameras contain an inching knob. This is a small knob, located either inside the camera body or on the outside of the camera. By turning this knob you are able to slowly advance or "inch" the film through the camera movement to check that it is moving smoothly. Whenever you thread the film into the camera, you should turn the inching knob a few turns to check that the film is traveling smoothly and not binding or catching anywhere.

Viewing System

The viewing system or viewfinder allows the camera operator to view the scene. There are three basic types of viewing systems that have been used on motion picture cameras over the years. The rack over viewing system and direct viewfinder are older viewing systems that are not used today for most professional motion picture productions. The current standard viewing system for professional motion picture cameras is the mirrored-shutter reflex viewfinder system. A reflex viewfinder is one that allows you to view the image directly through the lens, even during filming. The mirrored-shutter reflex system contains a mirror that is a part of the camera shutter. When the shutter is in the open position, all of the light entering the lens strikes the film and creates an exposure. When the shutter is closed, all of the light is directed to the eyepiece for the camera operator to view the shot (Figure 1.18).

Diopter Adjustment

Because of the differences in each person's eyesight, the viewfinder of most cameras has an adjustable diopter. By setting the diopter according to your particular vision, the image will appear in focus when you look through the eyepiece, provided the lens focus is set correctly. To adjust the diopter, it is best to remove the lens and point the camera at a bright light or white surface. While looking through the eyepiece, turn the diopter adjustment ring until the crosshair of the ground glass in the viewfinder is sharpest. A further discussion of the viewfinder adjustment is located in Chapter 4.

Motor

The three main types of camera motors are variable, constant, and crystal. Almost all professional motion picture cameras today use a crystal motor. The camera motor contains a crystal that is similar to

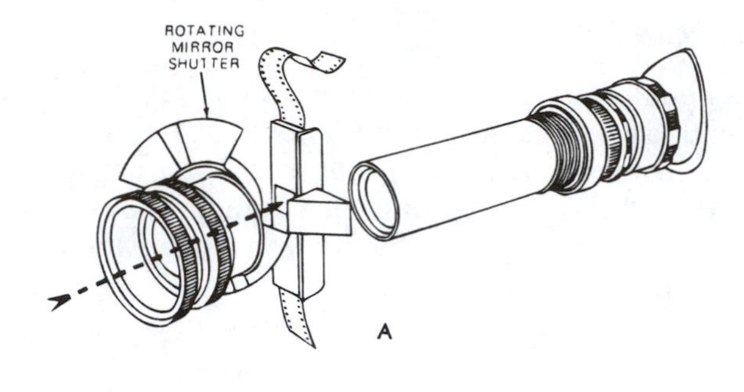

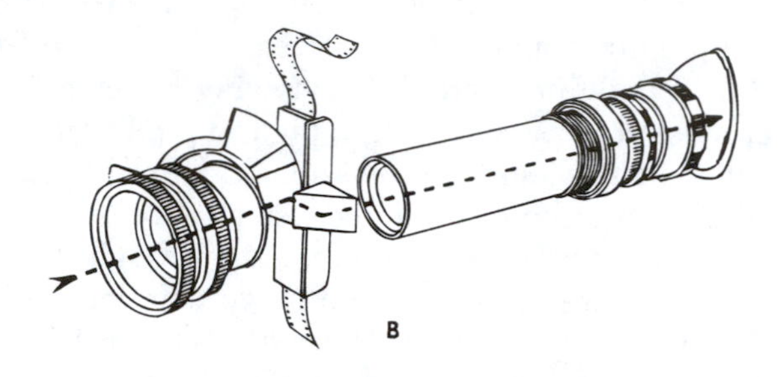

Figure 1.18 Illustration of a simple mirror reflex viewfinder system. (A) With mirror shutter open, all light is directed to the film. (B) With mirror shutter closed, all light is directed to the eyepiece. (Courtesy of Arriflex Corporation.)

the crystal found in a quartz watch. The sound recorder also contains a similar crystal. This crystal vibrates at a precise frequency, ensuring that during shooting the camera and sound recorder are running in sync so that the picture and sound will match. Most crystal motors have the ability to run at variable speeds for slow motion or high speed filming. They may also be set to various speeds other than sync speed of 24 f.p.s. by using some type of optional speed control device.

Lens

The lens is a device containing one or more pieces of optically transparent material, such as glass, which bends the rays of light passing through it, causing them to focus at a point. This point is called the *film plane* or *focal plane*, and the light causes an exposure on the film's emulsion at this point. All lenses are referred to by their focal length, and it is the focal length that determines the size of the image. The definition of *focal length* is the distance from the optical center of the lens to the film plane when the lens is focused at infinity. The *optical center* is a mathematical point within the lens that is determined when the lens is manufactured. The focal length of the lens is always measured in millimeters (mm).

When filming in the 35mm film format, a lens that has a focal length of 50mm is considered a normal lens because it approximates an image size that is the same as that seen by the human eye. When filming in the 35mm film format, as a general rule, any lens that has a focal length less than 50mm may be called *wide angle*, and any lens that has a focal length more than 50mm may be called *telephoto*. A wide angle lens will distort the image because it exaggerates distances and makes small rooms seem larger than they actually are. A telephoto lens compresses objects together and makes them appear closer than they actually are. In 16mm film format, a 25mm lens would be considered a normal lens. When filming in 16mm film format, as a general rule any lens that has a focal length less than 25mm may be called wide angle, and any lens that has a focal length more than 25mm may be called telephoto.

Primes and Zooms

When referring to lenses we usually speak of two main types. *Prime lenses* are lenses that are of a single, fixed focal length that cannot be changed. Prime lenses are referred to by their focal length, such as 25mm, 32mm, etc. *Zoom lenses* are lenses that have variable focal lengths that can be changed during shooting. By turning the barrel of the zoom lens, you are able to change the focal length. Zoom lenses are most often referred to by their range of focal lengths, such as 12mm to 120mm, 25mm to 250mm, etc. A further discussion of lenses can be found in Chapter 4.

Magazine

A *magazine* is a removable, lightproof container that is used to hold the film before and after exposure. Each camera uses a different type of film magazine. In any film magazine, the area that holds the fresh, unexposed raw stock is called the *feed side*. The area that holds the exposed film stock is called the *take-up side*. A further discussion of types of magazines and the procedure for loading and unloading them can be found in Chapter 3.

FILTERS

One of the most frequently used pieces of equipment in cinematography is the filter. It is a device that modifies the light reaching the film in order to achieve a specific effect. Filters may be placed in front of the lens, behind it, or even on the light source. For our purposes, we will deal only with filters that are placed on the camera, in front of the lens. Because some filters reduce the amount of light striking the film, an exposure compensation is sometimes required. The most common filters, their effect, and any exposure compensation will be briefly covered.

Conversion Filters

These are filters that are used to convert one color temperature to another. Because there are two different types of color balance for film (daylight or tungsten), there are two basic types of conversion filters.

85 Filter

When using tungsten-balanced film in daylight, a number 85 filter is used to correct for the difference in color temperature. When using this filter, an exposure compensation of two thirds of a stop is required. The 85 filter is orange or amber in color.

80A Filter

When using daylight-balanced film in tungsten light, a number 80A filter is used to correct for the difference in color temperature. When using this filter, an exposure compensation of two stops is required. The 80A filter is blue in color.

Neutral Density Filters

Many times when filming outdoors in daylight, the Director of Photography may wish to reduce the amount of light entering the lens, or reduce the depth of field for the shot. A neutral density filter would be used to do this. Neutral density filters are usually abbreviated ND.

The most commonly used neutral density or ND filters are ND3, ND6, and ND9. When using these, you must remember to adjust your exposure accordingly. The ND3 requires an exposure compensation of one stop, the ND6 two stops, and the ND9 three stops. For a complete discussion of depth of field, see Chapter 4.

Polarizing Filter

A polarizing filter reduces glare or reflections from shiny, nonmetallic surfaces such as glass and water. By placing the polarizer in front of the lens and rotating it, you are able to remove most of the unwanted reflections in a shot. Once the correct position of the polarizer has been determined, be sure to lock it in place so that it doesn't move during the shot. When using a polarizer filter, you must adjust your exposure by 1 1/2 to 2 stops.

Combination Filters

Any filter that combines two or more filters into one filter is called a *combination filter*. The most common combination filters are those that combine an 85 with the series of neutral density filters to get 85ND3, 85ND6, and 85ND9. Another common combination filter is the 85 plus Polarizer, which is usually called an 85POLA. When using a combination filter, in order to obtain the correct exposure compensation, you should add together the exposure compensation for each filter. For example, when using an 85ND6, your exposure compensation would be 2 2/3 stops—2 stops for the ND6 plus 2/3 of a stop for the 85.

Diffusion Filters

When talking about diffusion filters, we may be referring to many types and styles of filters that will give a similar effect. A general definition of a diffusion filter is a filter that is used to soften the image or look of the picture. A diffusion filter is usually made of glass containing a rippled surface, which prevents the light from focusing sharply. It will produce an image in which fine details are not clearly visible. It may also give the appearance that the image is out of focus. One of the most common uses of diffusion filters is to minimize or soften any facial blemishes or wrinkles on an actor or to help soften the image of an actor or actress who has many wrinkles in his or her skin. Some names of the most commonly used diffusion filters are Tiffen

Diffusion, Harrison & Harrison Diffusion, Black Dot Texture Screen, Black Pro-Mist, White Pro-Mist, Soft Net, Net, and Supa-Frost.

Two very popular types of diffusion filters being used are the Tiffen Black Pro-Mist and the White Pro-Mist. The White Pro-Mist softens the image without causing it to appear out of focus. It also spreads light slightly by creating a small amount of flare from light sources, and it will slightly reduce the contrast. The Black Pro-Mist softens the image with a more subtle flare from light sources and slightly reduces contrast by lightening shadows and darkening highlights.

Soft Nets and Nets actually are in a separate category called *nets.* A net may be any fine mesh type material that is placed on the camera and acts as a diffuser. In the early days of filmmaking, many Directors of Photography would stretch a stocking material over the front of the lens to create the diffusion effect. With the exception of the Black Dot Texture Screens, diffusion filters require no exposure compensation. All diffusion filters are available in sets ranging in density from very light to heavy diffusion.

Fog Filters and Double Fog Filters

In order to simulate the effects of fog, we use a fog filter or double fog filter. In a real fog situation, the fog causes lights to flare, and this is produced by the fog or double fog filter. The fog filters are also available in sets ranging in density from very light to very heavy. No exposure compensation is needed when using these filters.

Low Contrast Filters

In order to lighten the shadow areas of a scene, a director of photography may use a low contrast filter. This causes the light from the highlight areas of the scene to bleed into the shadow areas, which produces a lower contrast. In other words, it lightens the shadows without affecting the highlight areas. They do not soften the image or make it appear out of focus like diffusion filters do. As with diffusion and fog filters, they are available in varying densities. No exposure compensation is required for these filters.

Soft Contrast Filters

A soft contrast filter may be used to lower the contrast in a slightly different manner. It is different from the low contrast filter because it darkens highlights without affecting the shadow areas. They are also available in varying densities. No exposure compensation is required for these filters.

Ultra Contrast Filters

Another way to lower the overall contrast of a scene is to use the ultra contrast filter. This filter lowers the contrast evenly throughout the scene by equally lightening the shadow areas and darkening the highlight areas. They are available in varying densities and require no exposure compensation.

Coral Filters

In order to make a scene appear warmer, a Director of Photography may use a coral filter. The coral filter may be used when filming a sunset or a fireside scene in order to give the scene a warmer look. Another use of the coral filter is when filming outside in daylight. Because the color temperature of daylight changes from early morning to late afternoon, the Director of Photography may use a coral filter along with, or in place of, the 85 filter to give the scene a slightly warmer look. Coral filters come in varying densities, and an exposure compensation is required based on the density of the filter.

Enhancing Filters

In order to create brighter reds, oranges, and rust browns when filming, a Director of Photography may use the enhancing filter. While creating spectacular effects on the reds and oranges in the shot, it has very little effect on other colors. This filter is especially useful when filming fall foliage. The enhancing filter requires an exposure compensation of 1 $^{1}/_{2}$ to 2 stops.

Graduated Filters

Sometimes we only want to alter a portion of the frame with a specific filter. To do this the Director of Photography would use a graduated or grad filter. Only half of the filter contains the specific filter, while the remaining half is clear. For example, we may use a Neutral Density Grad or a Coral Grad for certain effects. Some graduated filters also come in varying densities.

Diopters

When doing extreme close-up work, we may need to use a special type of filter called a *diopter* on the lens. The diopter is actually a type of lens, but because it is mounted in front of a standard lens, similar to a filter, it is being mentioned here. The diopter allows the lens to focus closer than the lens normal focusing range allows. The diopters come in varying strengths as follows, +1/4, +1/2, +1, +1 1/2, +2, +3, etc. The higher the number, the closer the lens will focus. The glass of the diopter contains one side that is curved, and when placing the diopter on the lens this curved side must face away from the lens. When using a diopter, no exposure compensation is required.

Optical Flat

A special filter that should be included in every camera package is the optical flat. It is an optically corrected, clear piece of glass that has many uses. It may be placed in front of the lens to protect the lens for a shot in which something is being projected towards the camera. You also may be shooting in windy situations, where dust and dirt may be blown toward the lens, or if on a beach water may be blown towards the lens. The optical flat will protect the front element of the lens from these items. In addition, it can be used to cut down on the noise level of the camera. Much of the noise from the camera comes out through the lens. By placing an optical flat in front of the lens, you can reduce this noise and achieve a quiet sync sound take during shooting. So if the sound mixer asks you to place an optical flat on the camera, it usually means that he is hearing some camera noise through the microphone.

There are currently over 200 types of filters available for your use. The above-named filters are only a sampling of what is available to the cinematographer. This small listing is intended to give you a basic understanding of the most commonly used filters. Through experimentation and use of the filters, a Director of Photography usually knows which filter to use for a specific application or effect. For a listing of more filters and the common filter sizes, see Appendix B.

Filter Manufacturers

There are many different manufacturers of filters, including Tiffen, Harrison & Harrison, Mitchell, Wilson Film Services, and Fries Engineering.

CAMERA MOUNTS

There are many different devices and tools available to the cinematographer for mounting the camera, moving it, and keeping it smooth and steady when following the action within a scene.

Tripod and Spreader

One of the most common supports for the camera is a three-legged device called a *tripod*. Each of the three legs of the tripod can be adjusted in height according to the shot needed. The feet of the tripod are usually placed into an adjustable brace called a *spreader*. The spreader holds the legs in position and keeps them from collapsing when the legs are extended or spread out (Figure 1.19).

If a spreader is not available, many assistants use a piece of carpet approximately 4 feet by 4 feet in size. The points of the tripod feet will grip the carpet and prevent the tripod legs from moving or spreading apart. The two most commonly used tripods are the standard tripod and the baby tripod (Figure 1.20).

All tripods will have one of two types of top castings for the head to attach to. It will either be what is known as Mitchell flat base or bowl, or a ball base (Figure 1.21).

The two most commonly used names used to refer to the tripod are "sticks" and "legs."

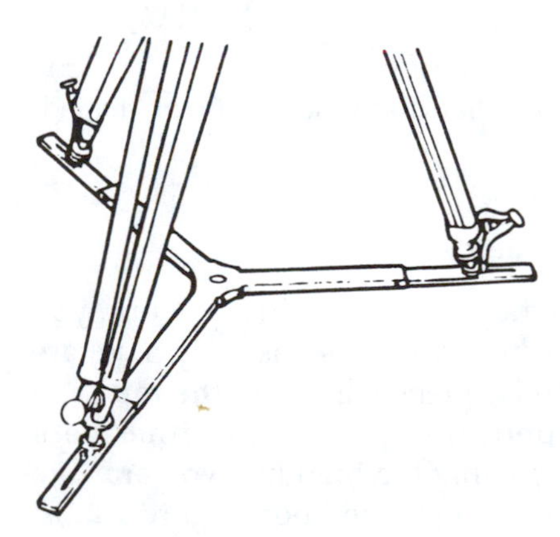

Figure 1.19
Tripod legs locked onto the spreader.

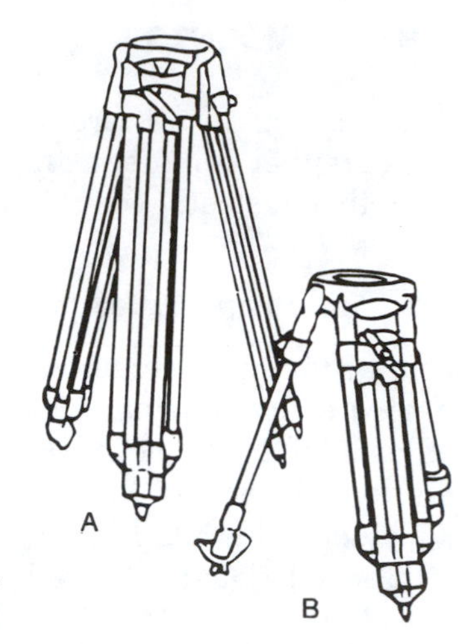

Figure 1.20
(A) Standard tripod. (B) Baby tripod.
(Courtesy of Panavision, Inc.)

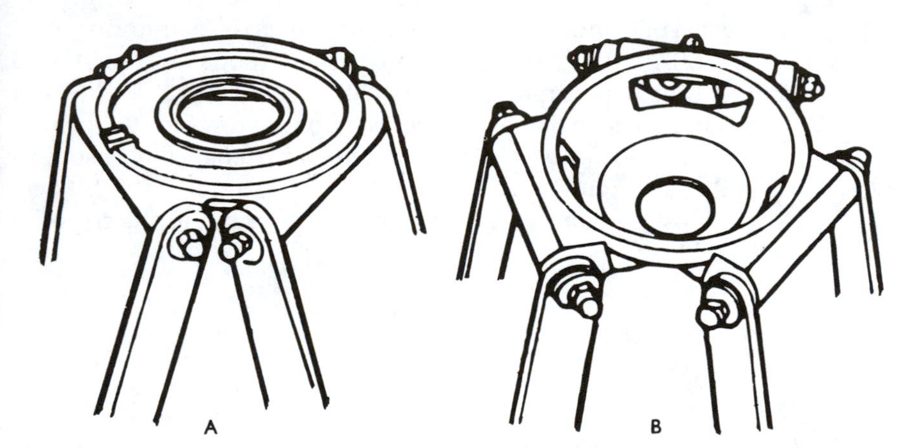

Figure 1.21 Tripod top castings. (A) Mitchell flat base. (B) Bowl shaped. (Courtesy of Panavision, Inc.)

High Hat

For doing extreme low angle shots where a tripod will not work, you will use a mounting device called a *high hat*. It is basically a square piece of wood onto which is mounted a piece similar to the top casting of the tripod. As with the tripod, this piece may either be a Mitchell flat base or bowl shaped. By using the high hat, you are able to get the camera lens just a few inches above the floor (Figure 1.22).

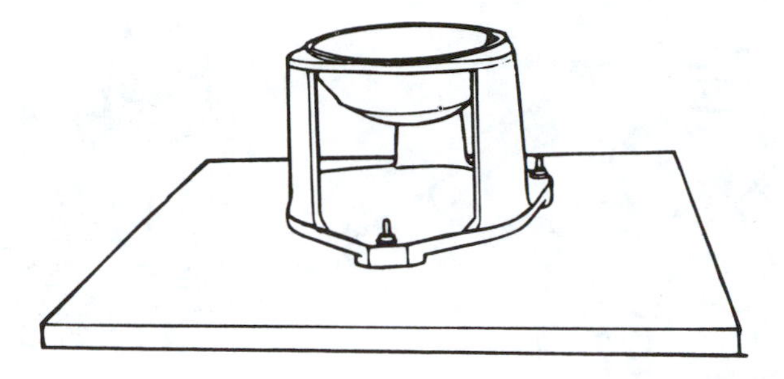

Figure 1.22 High hat with bowl-shaped top casting.

Tripod Heads

In order to make smooth moves with the camera in order to follow the action within a scene, the camera must be mounted onto some type of tripod head. This head will allow the camera operator to make smooth pan and tilt moves when following the action. Any horizontal movement of the camera to follow the action is called a *pan* or *panning*, and any vertical movement of the head is called a *tilt* or *tilting*. The two most common types of heads are the fluid head and geared head.

Fluid Head

The most commonly used tripod head, because of its ability to make smooth pan and tilt moves, is the fluid head. The internal elements of the head contain some type of viscous fluid, which provides a slight resistance against the movements. There is usually an adjustment on the outside of the head to increase or decrease the amount of resistance. Depending on the type of shot, the camera operator may want more or less resistance in order to make a smooth pan or tilt move. The pan and tilt movements are controlled by a handle, which is usually mounted to the right side of the head. By moving the handle left and right, or up and down, you are able to make smooth pan and tilt moves. Some of the most common fluid heads are manufactured by O'Connor, Sachtler, Vinten, Cartoni, Ronford Baker, and Weaver Steadman. When ordering a fluid head, be sure that it contains the same style base as the tripod top casting, either a Mitchell flat base or a ball leveling base (Figure 1.23).

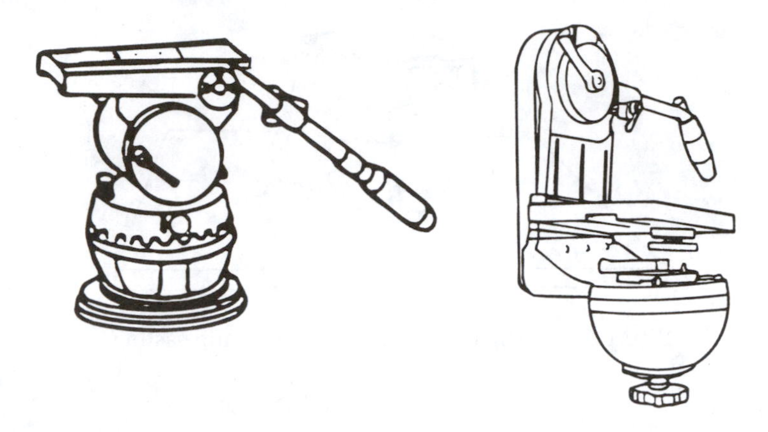

Figure 1.23 Two types of fluid heads.

Geared Head

For very precise and smooth movements, you might choose to use a geared head. The pan and tilt movements are controlled by two wheels, which are connected to gears or belts within the head. One of the wheels is located to the back of the head and it controls the tilt, and the other wheel is located on the left side and it controls the pan. It takes much practice to be able to operate the geared head correctly, but once you learn it you will most likely not want to use any other type of tripod head. Some of the most common geared heads are the Arriflex Arrihead, Panavision Panahead, Worral, and Mini Worral (Figure 1.24).

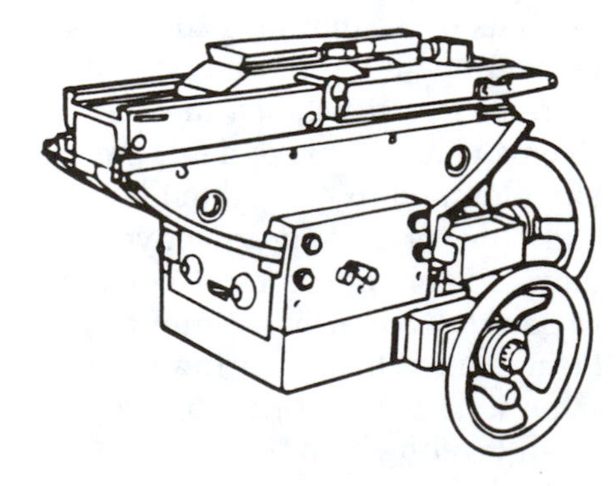

Figure 1.24
Panavision Panahead gear head.
(Courtesy of Panavision, Inc.)

Steadicam

A highly specialized mounting device for the camera is the steadicam. It is a body-mounted harness that is worn by the camera operator. It consists of a vest, a special support arm, and the basic steadicam unit onto which the camera is mounted. The arm consists of six springs, which absorb the up and down movement of the camera, allowing it to give a steady image. The steadicam allows the operator to do traveling shots where a dolly or crane is not practical, or to bring an actor from one location to another within the scene, without an edit. With the steadicam, the operator can follow an actor while running, up or down stairs or an incline, through a building, in wheelchair-mounted shots, car mount, and many other types of special shots. In order to be able to use the steadicam properly, an operator should attend special classes to be certified to use the system. Since the Steadicam was first introduced to the film industry, it has gone through many changes. There have been many different models of the system, including Model I, Model II, Model III, Model III-A, EFP, Master, and Pro (Figure 1.25).

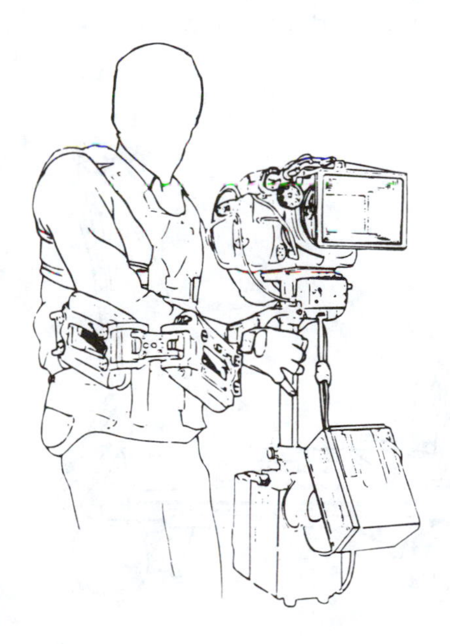

Figure 1.25
Steadicam operator using the system with an Arriflex 35BL camera. (Reprinted from the *Arri 35 Book*, with permission of the Arriflex Corporation.)

Dolly

A wheeled platform onto which the head and camera may be mounted is called a *dolly.* The dolly also has seats for the camera operator and camera assistant. Not all shots in a film are stationary. Some require the camera to move in order to follow the action within a scene. By mounting the camera to the dolly, you are able to do traveling or moving shots very smoothly. The dolly is usually placed on some type of track so that the movement will be free of vibrations. Most dollies contain some type of boom arm, which is operated hydraulically or by air pressure to raise or lower the height of the camera during a shot. When using a dolly, the same head that was mounted to the tripod may also be mounted to the dolly. The dolly contains a mounting platform similar to the top piece of the tripod so that the head may be locked firmly in place (Figure 1.26).

Crane

For many of the high angle shots that you may have seen in films, a piece of equipment known as a *crane* is usually used. The crane allows you to start a shot from a very high angle, and then boom down and move in to a very close shot. Or you may do the opposite, by starting low and then boom up to a high angle, wide shot. A crane should only be used when necessary. It is very important to have qualified

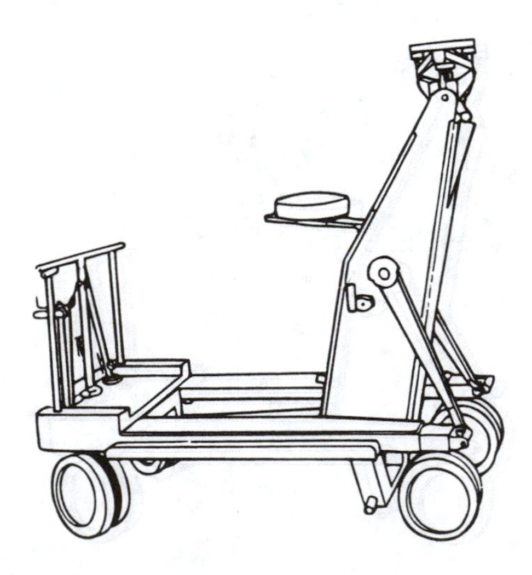

Figure 1.26
Camera dolly. (Courtesy of J.L. Fisher, Inc.)

grips who thoroughly know how to operate the crane safely. The most commonly used crane for motion pictures is manufactured by Chapman.

The above-mentioned pieces of equipment are only a small sampling of the wide variety of equipment used by a camera department. As you work more frequently on different types of productions, you will learn about and use many other specialized pieces of equipment. Whenever you come across a piece of equipment that you are not familiar with, ask the rental house to explain it to you so that you feel comfortable using it. Never try to use a piece of equipment that you are not familiar with.

2

The Camera Department

The number of members in the camera department will depend on the kind of film being shot. Big-budget feature films will usually have a larger crew than a low-budget film, commercial, or music video. In the United States the chain of command for the camera department is as follows:

Director of Photography (D.P.)

Camera Operator

First Assistant Cameraman (1st A.C.) (Focus Puller)

Second Assistant Cameraman (2nd A.C.) (Clapper/Loader)

Loader (optional position on larger multi-camera productions)

Each member of the camera department has specific duties and responsibilities, and each position is related to all the other positions. The D.P. is the head of the department. This chapter discusses the D.P. and the Camera Operator. Chapter 3 deals with the 2nd A.C., and Chapter 4 with the 1st A.C.

DIRECTOR OF PHOTOGRAPHY (D.P.)

The D.P. is the head of all technical departments on a film crew and is responsible for establishing how the script is translated into visual images based on the director's request. The D.P. will decide which camera, lenses, and film stock will be used for the production. The D.P. will hire the Camera Operator and probably also hire the 1st A.C. In many cases the D.P. will act as Camera Operator so that position will not be a part of the film crew. In hiring the 1st A.C., the D.P. usually bases the decision on past work experience and chooses someone he is comfortable working with. If the person he wishes to hire is not available, he may ask for a recommendation from his usual 1st A.C. or a recommendation from another D.P. The position of 1st A.C. is a very important one, and the D.P. wants to have someone who can be trusted and is good at the job. Because the 1st A.C. works closest with the 2nd A.C., the 1st A.C. usually hires the 2nd A.C. Again, this is usually based on past working experiences, or it may be based on a recommendation of another trusted camera assistant.

During shooting, all members of the camera department must work closely together as a team to get the job done. The D.P. decides where the camera is placed for each shot and which lens is to be used. It is up to the camera assistants to get the camera set up each time and place all appropriate accessories on the camera for shooting. The D.P. decides how the lights are to be placed for each shot, and once the lights are set, he gives the 1st A.C. the correct exposure to be set on the lens for shooting. In addition to working closely with the camera assistants, the D.P. will help the Camera Operator decide the composition for the shot. The D.P. will also decide if there are to be any dolly moves for the shot and when they will take place.

Many D.P.s started their film careers as camera assistants, so they should know and understand the requirements of the job. They probably worked a few years each as 2nd A.C. and 1st A.C. Following that they may have been a Camera Operator for a few years before finally becoming a D.P. The length of time that is spent at each position is based on each person's individual circumstances.

There are also many D.P.s who arrived at the position without ever having been a camera assistant. They may have been a lighting technician or gaffer before becoming a D.P. They also could have started their career as a documentary filmmaker or television news cameraperson. If the D.P. has never been a camera assistant, he may not be fully aware of all the duties of the job. In any case you must be able to work closely with the D.P. in order to get the job done.

Many D.P.s started out working on small, low-budget films, or even some student film projects. These small projects enabled them to gain valuable experience that later helped them get their first big break on a major, large-budget production. Some directors of photography started out as apprentices to well-known D.P.s. By working with these professionals, they learned many valuable skills that helped them when it came time to start out on their own.

The following are some of the responsibilities of the D.P. They are listed in no particular order.

- works with the director, production designer, and set construction supervisor to determine the "look" of the film and how the sets will be designed and constructed
- assists the director in translating the screenplay into visual images
- attends production meetings to discuss the script and make any suggestions to help the production run smoothly
- goes on location scouts with the director and any other production personnel to help determine their suitability for filming
- supervises any camera tests that may be necessary
- supervises any lighting, costume, and makeup tests
- maintains the photographic quality of the production
- hires the members of the camera crew, the gaffer, and the key grip
- chooses camera, lenses, filters, film stock, and any other camera equipment that may be needed
- works with the grip and electric crews to determine the type and quantity of equipment needed for each department
- sets the camera position, camera angle, and any camera movement for each shot based on the Director's request
- works with the Director when lining up and matching action and screen direction from shot to shot
- determines the correct exposure (f-stop) for each shot
- selects the lens for each shot
- works with the Camera Operator to set the composition for each shot based on the Director's request
- maintains the continuity of lighting from scene to scene
- supervises the crews for all cameras in use on the production
- supervises each technical crew while on stage or location
- views dailies with the director and other production personnel
- supervises the timing of the final print of the film
- supervises the transfer from film to videotape
- provides exposure meters and other necessary tools associated with performing the job

CAMERA OPERATOR

The next person in line in the camera department is the Camera Operator. In the United States, the Camera Operator works closely with the D.P. to determine the composition for each shot as instructed by the director. In Britain, the Director and the Camera Operator work together to determine the placement of the camera and the composition of the shots. The D.P., or Lighting Cameraman as he is sometimes called, deals primarily with the lighting of the set.

The primary job of the Camera Operator is to make smooth pan and tilt moves in order to maintain the composition of the subject. The Camera Operator keeps the action within the frame lines in order to tell the story. Sometimes the Camera Operator decides the placement of the camera for each shot and also the lens that will be used. The 1st A.C. works closest with the Camera Operator during rehearsals and actual shooting. There may be a complicated camera move that requires zoom lens moves and many focus changes during the shot. The Camera Operator rehearses these moves with the 1st A.C. before shooting them to be sure they are done at the correct time during the shot. If a problem arises with any of these moves during the shot, the Camera Operator is the only one who can detect it and must let the 1st A.C. know where the problem occurred so that it can be corrected for the next shot.

The Camera Operator rehearses any dolly moves that may have been determined by the D.P. The Camera Operator lets the dolly grip know when it is the right time to move the dolly during the shot. The Camera Operator also works with the sound department boom operator to set the placement of the boom microphone during the shot. He may let the boom operator look through the camera to see the frame size, or he may just tell him where the edge of frame is so that the microphone is placed where it is not in the shot. The Camera Operator also tells the 2nd A.C. if any actor's marks are visible in the frame and if they should be made smaller for the shot. When actual shooting starts, the Camera Operator sometimes instructs the 2nd A.C. where to place the slate so that it is visible in the frame.

The following are some of the responsibilities of the Camera Operator. They are listed in no particular order.

- adjusts the viewfinder eyepiece for her vision
- maintains the composition as instructed by the director or the D.P.
- watches to make sure the proper eye lines and screen directions are maintained
- makes smooth pans and tilts during each shot to maintain the proper composition

- approves or disapproves each take after it is shot
- works closely with the 2nd A.C. regarding the proper size and placement of actor's marks; if the marks are seen in the shot, the Camera Operator informs the 2nd A.C. to make them smaller or to remove them
- notifies the 2nd A.C. when the camera has reached sync speed so that he may slate the shot
- works closely with the 1st A.C. during rehearsals and takes to ensure proper focus and zoom moves
- works closely with the dolly grip during rehearsals and takes to ensure smooth dolly or crane moves
- works closely with the sound department to ensure the proper placement of microphones during each take
- may act as D.P. on any second unit shooting during the production
- views dailies with the Director and other production personnel

The responsibilities of the First Assistant Cameraman and Second Assistant Cameraman are covered in detail in Chapters 3 and 4. Listed below is a brief summary of the responsibilities of each of these positions. They are listed in no particular order.

FIRST ASSISTANT CAMERAMAN (1ST A.C.)

- works with the D.P. and/or Camera Operator to choose the camera equipment that will be used on the production
- works with the 2nd A.C. to prepare a list of expendables, which is given to the production office
- preps the camera package alone or along with the 2nd A.C.
- mounts the camera head onto the tripod, dolly, or other support piece and ensures that it is working properly
- unpacks and assembles the camera and all of its components at the start of each shooting day
- loads and unloads film into the camera
- resets the footage counter to zero after each reload
- keeps all parts of the camera clean and free from dirt and dust; camera body, lenses, filters, magazines, and so on
- oils and lubes the camera as needed
- sets the viewfinder eyepiece for each key person looking through the camera
- prior to each shot, ensures that the camera is level

- if the camera is mounted on a tripod, ensures that it is securely positioned and leveled
- checks to be sure that no lights are kicking into the lens, causing a flare, when the camera is in its proper position
- places proper lens, filter, and any other accessory on the camera as instructed by the D.P. or Camera Operator
- sets the t-stop on the lens prior to each take as instructed by the D.P.
- measures the distances to subjects during rehearsals and marks the lens or focus marking disk
- follows focus and makes zoom lens moves during takes
- adjusts the shutter angle, t-stop, or camera speed during a take, as needed and as instructed by the D.P.
- gives the 2nd A.C. footage readings from the camera after each take
- after each printed take, checks the gate for hairs or emulsion buildup and requests another take if necessary
- supervises the transportation and moving of all camera equipment between filming locations
- works with the 2nd A.C. to move the camera to each new position
- if there is no 2nd A.C. on the production, then also performs those duties
- disassembles the camera and its components at the completion of the shooting day and packs them away into the appropriate cases
- at the completion of filming, wraps and cleans all camera equipment for returning to the rental house
- provides all the necessary tools and accessories associated with performing the job

SECOND ASSISTANT CAMERAMAN (2ND A.C.)

- before production obtains a supply of empty cans, black bags, camera reports, and cores from the lab
- prepares a list of expendables with the 1st A.C.
- preps the camera package along with the 1st A.C.
- cleans the camera truck and/or darkroom for use during the production, and makes sure that each is loaded with the proper supplies and equipment
- communicates with the script supervisor in order to obtain the scene and take number for each shot, and also which takes are to be printed

- records all information on the slate
- records all information on the camera reports
- loads and unloads film in the magazines, and places proper identification on each
- helps to set up the camera at the start of each shooting day
- marks the position of actors during the rehearsals
- slates each scene, whether sound (SYNC) or silent (MOS)
- assists in changing lenses, filters, magazines, and so on, and in moving the camera to each new position
- sets up and moves video monitor for each new camera setup
- prepares exposed film for delivery to the lab and delivers it to the production company representative at the end of each shooting day
- maintains a record of all film received, film shot, short ends created, and film on hand at the end of each shooting day during the production
- maintains an inventory of film stock and expendables on hand and requests additional supplies from the production office as needed
- distributes copies of the camera reports and film inventory forms to the appropriate departments
- keeps a file of all paperwork relating to the camera department during the production
- performs the job of 1st A.C. if necessary
- at the end of each shooting day, helps the 1st A.C. pack away all camera equipment in a safe place
- at the completion of filming, helps the 1st A.C. wrap and clean all camera equipment for returning to the rental house
- at the completion of filming, cleans and wraps the camera truck
- provides all the necessary tools and accessories associated with performing the job

3

Second Camera Assistant

In most cases, when you first join the camera department, you will be starting as a Second Camera Assistant, or 2nd A.C. In Britain and Europe, the 2nd A.C. may be called the Clapper/Loader. Sometimes you may start as a Loader, which is very similar to the 2nd A.C. The Loader is primarily responsible for loading and unloading film into the magazines, and filling in all of the paperwork for the camera department. The Loader usually never leaves the camera truck or loading area. There may be some occasions when he works alongside the 2nd A.C. on the set in order to gain further experience.

Many of the job duties of the Loader are the same as for a 2nd A.C., so if you are working as a Loader, most of this chapter will also apply to you. The main difference between the Loader and 2nd A.C. is that the 2nd A.C. has more responsibilities. The 2nd A.C. works directly with the First Camera Assistant (1st A.C.) during the production and performs many different job duties each shooting day. This chapter discusses in detail each of the 2nd A.C.'s duties and responsibilities. Since there are three different stages of production, these duties are separated into three categories: pre-production, production, and post-production.

PRE-PRODUCTION

Obtaining Supplies

Once you have been hired for the job, you should find out which lab will be processing the film during the production. You should then either go to the lab and pick up a supply of empty film cans, black bags, camera reports, and spare cores, or ask the production company to arrange to have these items picked up. Remember to obtain various sized cans and bags. These items are necessary so that you can do the job properly, and you must have them available to you during production. The black bag is usually made out of paper or some type of plastic material. It is used to protect the roll of film from light and also from scratches when it is placed in the film can. You should never place a roll of film into a film can without first placing it in a black bag. Many assistants may also refer to the changing bag as a "black bag." It is not the same as the black bags used for wrapping exposed and unexposed film in a film can. The cans and bags will be used to *can out*, which means to wrap and store any short ends and any exposed film during the production.

The spare cores are needed to wind the film on the take-up side of the magazines, if the magazine does not have a collapsible core. The camera reports will be filled out during shooting. Many times the production company will have already picked up these supplies for you, but it is a good idea to have a supply of your own in case of emergency. Keep a constant inventory of these items as you should not run out of any of them during shooting. As shooting progresses you may ask the production office to have someone pick up additional supplies as needed. Never wait until you have run out before ordering additional supplies. It is better to have extra supplies on hand than to run out at a critical time during shooting.

Remember: Never run out of film cans and black bags.

Choosing Expendables

During pre-production the 2nd A.C. and 1st A.C. will prepare a list of expendables. This list will then be given to the production office so that they may purchase these items for the camera department. The expendables are items that will be needed daily in the performance of your job, such as camera tape, permanent felt tip markers, ballpoint pens, compressed air, lens tissue, lens cleaning solution, and so on. They are referred to as *expendables* because they are items that are used up or expended during the course of the production. It is usually a good idea for both assistants to prepare the list because each may need specialty items that should be included with the basic supplies. In addition, you should check with the D.P. and Camera Operator to see if there are any special items that they may need. The first order should give you enough supplies to start filming, and as the shooting progresses, check the expendable supply daily or weekly to see if anything is getting low and if you need to order more. When you see that additional items are needed, prepare a list and present it to the production office so that they may purchase the items for you. As with lab supplies, do not wait until you run out to order expendables. For a complete list of the standard items on a camera department expendables list, see Appendix C.

Remember: Never run out of expendables that you may need to do the job.

Preparation of Camera Equipment

Camera preparation, or prep, is usually done by the 1st A.C., but many times on larger productions the 2nd A.C. also works on the camera prep. Please see the section on camera prep in Chapter 4 for the procedures to follow.

Preparation of Camera Truck

When the camera prep has been completed, if you are using a camera truck, the equipment should then be loaded onto the truck. Before loading the truck, be sure that it has been cleaned out. Sweep the floor and clean off the shelves. If the truck is kept clean, there is less chance of the camera and equipment getting dirty.

Once the truck has been cleaned, load the equipment on the shelves. The shelves should then be labeled as to what is on each of them. When loading the camera truck, common sense is the key. Do not place camera, lenses, or filters on high shelves, where they may fall while the truck is moving. These items should be kept on lower shelves for safety. The camera case is usually kept under the workbench so that it may be accessed easily each shooting day. All shelves should have some type of straps across them to prevent cases from sliding off during moving. By using a logical system and order as to how the truck is loaded, you will be able to quickly set up at the start of each day and locate any item in a hurry. If the truck is equipped with a darkroom, it should be cleaned and stocked with all necessary supplies. Figure 3.1 shows a typical camera truck setup.

Figure 3.1 Typical camera truck setup. (Reprinted from the *Arri 35 Book*, with permission of the Arriflex Corporation.)

Preparation of Darkroom

When using a darkroom, whether it is on a stage or on a camera truck, the first thing you should do is to be sure that it is lightproof. The best way to do this is to go into the darkroom, close the door, turn off the light, and stay in for at least 5 minutes to allow your eyes to adjust to the dark. After approximately 5 minutes, hold your hand 12 inches from your face. If you are able to see your hand, then light is leaking in. Find the leaks and plug them or cover them with tape. Check along the floor, walls, and ceiling, and along the door frame where it closes. Never use a darkroom until you are sure that it is completely lightproof. Also, be sure that the door has a lock on the inside to prevent anyone from opening it while you are loading or unloading film. The darkroom should be checked daily to be sure that no light is leaking in. The darkroom should especially be checked daily if it is located on a camera truck that has been driven from location to location. The shifting and swaying of the truck during movement can cause the seams of the walls, floor, and door of the darkroom to separate.

Once you are sure that the darkroom is lightproof, clean and stock it with all necessary supplies and equipment. Only the items that are needed for the loading and unloading process should be kept in the darkroom. Any additional items may be stored on other shelves in the camera truck so they do not clutter the darkroom.

Set up the darkroom in a neat and orderly manner, with each item having an assigned location. This will help you do the job much faster so you do not have to search for something each time you load or unload. Camera tape, pens, permanent felt tip markers, compressed air cans, empty cans, and camera reports should all be kept within easy reach. Always separate the raw stock and short ends from the exposed film. Raw stock is any fresh, unexposed film, and short ends are short, unexposed rolls of film left over from a full roll. See Figure 3.2 for a typical darkroom setup.

PRODUCTION

Once you have completed all of the pre-production procedures, it is time for filming to start. The production phase of shooting is a complex operation that requires a great deal of dedication and hard work on the part of all involved, especially the 2nd A.C. The proper performance of the duties and responsibilities of the 2nd A.C. are vital to the

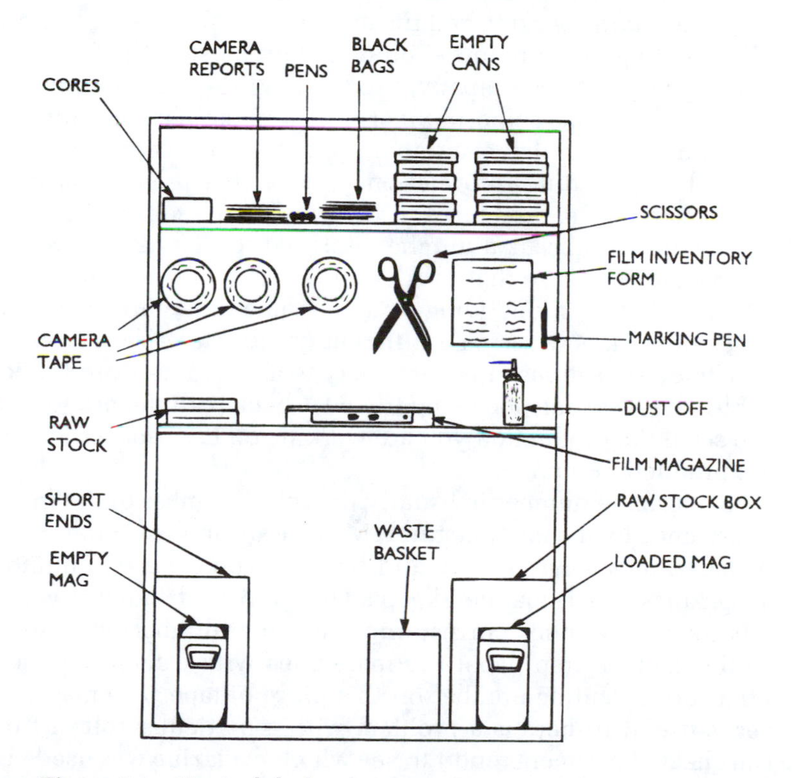

Figure 3.2 Typical darkroom setup.

smooth operation of the production. You must keep track of how much film is shot, how many rolls are used, which scenes and how many takes of each are shot, along with many other aspects of the job. You should be very organized and be able to jump in at a moment's notice with any information or equipment needed during shooting.

Camera Reports

Each roll of film shot during the production should have a camera report that shows which scenes were shot and how much film was used for each shot. Each lab has its own style of camera report, and each one contains the same or similar basic information. It is usually a good idea to use the report from the lab that will be processing the film, but if it is not possible, then any camera report will be all right. No matter what style of camera report you use, the basic information on it is the same. I separate all camera reports into two sections, one that I call the *heading section* and the other that I call the *shooting section*. The heading of the report should contain most of the following information: production company, production title, production number, Director, D.P., magazine number, roll number, camera number, footage, film type, emulsion number, date, and developing instructions. The basic heading information, such as the production company, production title, director, and D.P., should be self-explanatory. The production number is a number assigned to that particular production by the company that is filming it. The company may have many different productions going on at the same time, and one way to keep track of them is to assign a different number to each one. When filming a television series, it is customary to assign a new production number to each new episode being filmed. Check with the production office to see if the production you are working on has been assigned a specific production number.

The magazine number is usually the serial number of the magazine as assigned by the manufacturer. Many assistants assign numbers to the magazine, such as 1, 2, 3, and so on. If you choose to number your magazines in this manner, keep a written record showing which magazine serial number corresponds to your numbering system. During the camera prep, label the magazines with camera tape and place the corresponding number on this piece of tape. The magazine number is useful if there is a problem with a particular roll of film. You can check the camera report to see which magazine was used and have it repaired if necessary. The roll number is assigned each time the camera is loaded with a new roll of film. The usual procedure is that the first roll of film placed on the camera on the first day of shooting is roll number one (1), the next one is roll number two (2), then roll number three (3), and so on. Each time a new roll is placed on the camera, it is assigned a new number, whether it is a full roll of film or a short end. On each new shooting day, the roll number that you start with will usually be the next higher number from the one you ended with on the previous day. For example, on day number ten (10) of shooting, you ended with roll number forty-seven (47). When you start day number eleven (11), the first roll placed on the camera will be roll number forty-eight (48). The exception to this is if you do not remove the last roll of film from the camera on the previous day, and continue with it on the next day. If more than one camera is being used, it is standard to make the roll number a combination of the camera letter and the roll number, such as A-1, A-2, B-1, B-2, and so on. It is a good idea to check with the editors or the production company to see how they would like the roll numbers labeled each day.

The camera number is actually a letter assigned to the camera during camera prep. If only one camera is being used, then no number is assigned, but if you are using more than one camera, then the primary camera would be "A," the second camera would be "B," then "C," and so on. The footage refers to the amount of footage loaded into the magazine corresponding to the camera report. It is not always the same as the size of the magazine. Many times a short end will be loaded into a magazine instead of a full roll of film. The film type refers to what film stock you are using, for example, Kodak 7248, 7293, 5245, 5287, 5296; Fuji 8621, 8631, 8551, 8571; and so on. The emulsion number is the emulsion number and roll number assigned by the manufacturer. The film type and emulsion number can usually be obtained from the film can label. For example, if you are using Eastman Kodak Color Negative 7293-032-1902, the film type is 7293 and the emulsion number is 032-1902, and for Eastman Kodak Color Negative 5296-197-1102, the film type is 5296 and the emulsion number is 197-1102 (Figure 3.3).

You will notice on the breakdown of the information on the Kodak film can label, for film stock 5296-197-1102, the breakdown only lists 197 as the emulsion number and 1102 as the roll number. Since the film is manufactured in large batches, this information means that it is roll number 1102 of emulsion number 197. When filling in the camera report, you should always include both numbers when referring to the emulsion number.

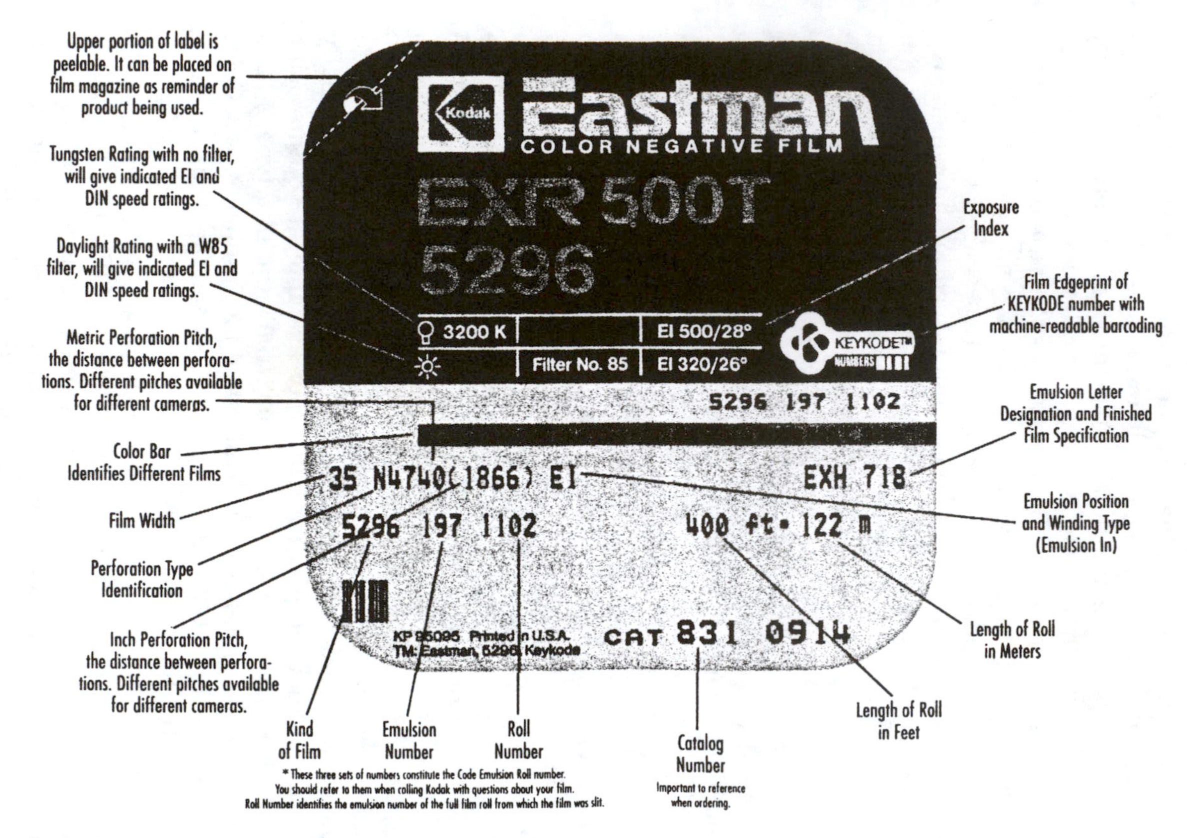

Figure 3.3 How to read the Kodak film can label. (Courtesy of Eastman Kodak Company.)

The date corresponds to the date that the roll of film is exposed. The developing instructions will usually be given to you by the D.P. In most cases it would be "DEVELOP NORMAL."

Because the basic heading information, such as production company, production title, Director, and Director of Photography, will remain the same during the production, it may be filled in ahead of time in order to save time. Many times, the film type and emulsion number and footage amount may also be filled in ahead of time if it is the same for the entire production. During shooting you will fill in the shooting portion of the camera report with the following information: scene number, take number, dial reading, footage, remarks, G (good), NG (no good), W (waste), T (total), and SE (short end).

Figures 3.4, 3.6, and 3.8 illustrate examples of the different styles of camera reports. Each one of these different styles will be discussed separately. Two examples of each camera report style, one blank and one completely filled in, are shown in this section so that you can compare the differences.

During shooting you will receive the scene number and take number from the Script Supervisor. Write these numbers in the appropriate space on the report. At the end of each take, you will check the footage counter on the camera to obtain the dial reading. If you cannot see the footage counter, ask the 1st A.C. to give you the information. Many times the 1st A.C. calls out the reading to you at the end of each take or gives you a hand signal to indicate the number on the camera footage counter. (See Chapter 4 for more information on the hand signals.) The dial reading on the camera report will be a form of the footage amount shown on the camera footage counter. Most professional motion picture cameras have some type of digital footage counter, and we assume that is the case here. When a new roll is placed on the camera, the footage counter should be reset to zero. Each time the camera is turned on, the numbers on the footage counter get progressively higher. To make the addition and subtraction on the camera report easier, round all dial readings to the nearest ten. As we all should have learned in elementary arithmetic, if the number ends in 0, 1, 2, 3, or 4, round it down, and if it ends in 5, 6, 7, 8, or 9, round it up. For example, if the camera footage counter shows a reading of 247, round it to 250. On the camera report, next to the appropriate scene and take number, write the number 250 in the dial column.

To determine the footage amount for each take, you subtract the previous dial reading from the one just recorded. For example, if the previous dial reading on the camera report is 210 and the present dial

Table 3.1 Camera Footage Counter Readings and Corresponding Camera Report Dial and Footage Amounts

CAMERA FOOTAGE COUNTER	CAMERA REPORT DIAL	CAMERA REPORT FOOTAGE
66	70	70
121	120	50
162	160	40
205	210	50
247	250	40
279	280	30
364	360	80
433	430	70
498	500	70
550	550	50
607	610	60
649	650	40
703	700	50
754	750	50
802	800	50
836	840	40
885	890	50
942	940	50
968	970	30

reading is 250, the footage for the present take is 40, 250 – 210. Table 3.1 shows an example of camera footage counter amounts and the corresponding dial reading and footage amounts for each. The information in Table 3.1 is used in the three different styles of camera reports so that you can compare the differences between each report.

The first camera report style is shown in Figures 3.4 and 3.5.

The SD column may be used to indicate whether the scene was shot sync or MOS. If the shot was done with sync sound, write S in the column for sync, and if it was done without sound, write M in the column for MOS. Most assistants do not use this column on the camera

Company
Pic. Title Prod. No.
Director
Cameraman Cam. No.

CAMERA REPORT 47200
Customer Order No.
Sheet No. of

Mag. No. Roll No.
Footage Date Exposed
Film type Emulsion No.
Develop ☐ Normal ☐ Other

FOTO-KEM Industries, Inc.
2800 West Olive Avenue
Burbank, California 91505
(818) 846-3101 FAX (818) 841-2120

SCENE NO.	TAKE	DIAL	FEET	SD.	REMARKS	SCENE NO.	TAKE	DIAL	FEET	SD.	REMARKS
										G	
										NG	
										W	
										T	

FK 13-3 (R 9/91)
40M 7/92 - 35141

THIS CAMERA REPORT SUBJECT TO PROVISIONS ON THE REVERSE SIDE

Figure 3.4 Example of one camera report style. (Courtesy of Foto-Kem Industries, Inc.)

report. In the Remarks column of the report, you may record a variety of information, including filters used, f-stop, focal length of lens, camera to subject distance, MOS, if the shot was done without sound, or tail slate or second slate, depending on the situation. You also may make a note whether the shot was interior (int), exterior (ext), day (day), or night (nite). There is no set rule as to what information should go in the Remarks column. Check with the D.P. and 1st A.C. to see if they want anything written in this space. Each production will be different. An example of this type of camera report completely filled out is shown in Figure 3.5.

Company	Demi Monde Productions		
Pic. Title	"Claire of the Moon"		
Director	N. Conn		
Cameraman	R. Sellars		
Mag. No.	10146	Roll No.	5
Footage	1000'	Date Exposed	10/29/91
Film Type	5298	Emulsion No.	237 - 4862
Develop	☒ Normal	☐ Other	

SCENE NO.	TAKE	DIAL	FEET	SD	REMARKS	SCENE NO.	TAKE	DIAL	FEET	SD	REMARKS
54	1	70	70			82 B	(1)	800	(50)		
	(2)	120	(50)				2	840	40		
	3	160	40			36	(1)	890	(50)		
	(4)	210	(50)				(2)	940	(50)		
54 A	(1)	250	(40)				3	970	30		
	2	280	30								
82	(1)	360	(80)				OUT AT 970'				
	2	430	70								
	(3)	500	(70)								
	4	550	50			DEVELOP NORMAL					G 600
	(5)	610	(60)			1 - LITE PRINT					NG 370
82 A	1	650	40								W 30
	(2)	700	(50)								T 1000
	(3)	750	(50)								

Figure 3.5 Example of completed camera report from Figure 3.4.

For the type of camera report shown in Figure 3.6, write the scene and take numbers as you did in the previous style of report.

Round the dial reading and put it in the Dial or Counter column, depending on which type of report you are using. In the Print column, write the footage only for the takes that are to be printed. Put the same information in the Remarks column as in the previous example. An example of this type of camera report completely filled out is shown in Figure 3.7.

As you can tell from looking at the third type of camera report shown in Figure 3.8, the only sections that are the same are the Scene number and the Remarks column.

Instead of writing down the dial readings in one column and the footage amounts in another column, only the footage amount is written in the space for the particular take number. Column 1^5 is for take 1 and for take 5, column 2^6 is for take 2 and take 6, and so on. Because

CAMERA REPORT

Technicolor® 28245

4050 LANKERSHIM BLVD.
N. HOLLYWOOD, CALIF. 91608

(818) 769-8500 Telex 674108

PROD. COMPANY/#

FILM TITLE DATE

CAMERAMAN DIRECTOR

CAMERA I.D. MAG. NO. ROLL #

☐ 35 mm. ☐ Color ☐ One Light
☐ 16 mm. ☐ B & W ☐ Timed

Type of Film/Emulsion

Processing — **Normal** ☐ **Forced 1 Stop** ☐ **Forced 2 Stop** ☐

SCENE NO.	TAKE	DIAL	PRINT	DAY / NIGHT	INT. / EXT.	REMARKS
						FOOTAGE
						GOOD
						N.G.
						WASTE
						TOTAL

Form 756 Technicolor is a Registered Trademark

Figure 3.6 Example of a second camera report style. (Courtesy of Technicolor.)

PROD. COMPANY Demi Monde Productions		
FILM TITLE "Claire of the Moon"	DATE 10/29/91	
CAMERAMAN R. Sellars	DIRECTOR N. Conn	
CAMERA I.D. A	MAG NO. 10146	ROLL # 5
☒ 35mm	☒ Color	☒ One Light
☐ 16mm	☐ B & W	☐ Timed
Type of Film/Emulsion 5298 - 237 - 4862		1000'
Processing - Normal ☒	Forced 1 Stop ☐	Forced 2 Stops ☐

SCENE NO.	TAKE	DIAL	PRINT	REMARKS
54	1	70		
	(2)	120	(50)	
	3	160		
	(4)	210	(50)	
54 A	(1)	250	(40)	
	2	280		
82	(1)	360	(80)	
	2	430		
	(3)	500	(70)	
	4	550		
	(5)	610	(60)	
82 A	1	650		
	(2)	700	(50)	
	(3)	750	(50)	
82 B	(1)	800	(50)	
	2	840		
36	(1)	890	(50)	FOOTAGE
	(2)	940	(50)	GOOD 600
	3	970		N.G. 370
				WASTE 30
	OUT AT 970'			TOTAL 1000
DEVELOP NORMAL		1 - LITE PRINT		

Figure 3.7 Example of completed camera report from Figure 3.6.

there is no space for the dial readings, you should write them along the left or right edge of the report just as a reference. An example of this type of camera report completely filled out is shown in Figure 3.9.

So, for this camera report style, scene 54, take 1, was 70 feet, take 2 was 50 feet, take 3 was 40 feet, and so on.

You will notice on each of the above styles of camera reports, certain take number and footage amounts have circles drawn around them. After each setup, the Script Supervisor will tell you which takes are to be circled. These are the takes that the Director wants to use for editing the film, and they are called the *good* or *printed takes*. Circling

deluxe laboratories, inc.

1377 NORTH SERRANO AVE., HOLLYWOOD, CA 90027 • PHONE 462-6171

color by deluxe®

camera report
sound report

No. 11675

DATE ______________ CUSTOMER ORDER NUMBER ______________

COMPANY ______________

DIRECTOR ______________ CAMERAMAN RECORDIST ______________

PRODUCTION NUMBER OR TITLE ______________

MAGAZINE NUMBER ______________ ROLL NUMBER ______________

TYPE OF FILM AND TYPE OF DAILIES

PLEASE CIRCLE

35 MM COLOR	35 MM B/W	16 MM COLOR	16 MM B/W
ONE LIGHT	ONE LIGHT	ONE LIGHT	ONE LIGHT
TIMED DAILIES	TIMED DAILIES	TIMED DAILIES	TIMED DAILIES

FORCE DEVELOP: ONE STOP TWO STOPS

TYPE OF FILM / EMULSION ______________

PRINT CIRCLED TAKES ONLY:

SCENE NUMBER	TAKES				REMARKS	
	1	2	3	4	DAY OR NIGHT	INTERIOR OR EXTERIOR

TOTAL FOOTAGE ______________

All contracts with this company are accepted with the understanding that all film delivered to it is covered by the owner against loss. This company takes every necessary precaution for the safekeeping of the film, but assumes no responsibility for its loss.

DEL 82 Rev. 7-76

Figure 3.8 Example of a third camera report style. (Courtesy of Deluxe Laboratories, Inc.)

DATE 10/29/91			
COMPANY Demi Monde Productions			
DIRECTOR N. Conn		CAMERAMAN R. Sellars	
PRODUCTION TITLE "Claire of the Moon"			
MAGAZINE NO. 10146		ROLL NO 5	
TYPE OF FILM AND TYPE OF DAILIES PLEASE CIRCLE			
35MM COLOR	35MM B/W	16MM COLOR	16MM B/W
ONE LIGHT	ONE LIGHT	ONE LIGHT	ONE LIGHT
TIMED DAILIES	TIMED DAILIES	TIMED DAILIES	TIMED DAILIES
FORCE DEVELOP:	ONE STOP	TWO STOPS	
TYPE OF FILM/EMULSION 5298 - 237 - 4862 1000'			
PRINT CIRCLE TAKES ONLY:			

	TAKES				
SCENE NO.	1[5]	2[6]	3[7]	4[8]	REMARKS
54	70	(50)	40	(50)	70 120
54 A	(40)	30			160 210
82	(80)	70	(70)	50	250 280
	(60)				360 430
82 A	40	(50)	(50)		500 550
82 B	(50)	40			610 650
36	(50)	(50)	30		700 750
					800 840
	OUT AT 970				890 940
					970
	DEVELOP NORMAL				G 600
	1 - LITE PRINT				NG 370
					W 30
					T 1000

Figure 3.9
Example of completed camera report from Figure 3.8.

lets the lab know which takes are to be printed during processing. If the film is being transferred to videotape, the circled takes are the transferred ones. Most labs will not print circled takes in 16mm because it is cheaper to print the entire roll, but for a video transfer, circled takes are used in 16mm. When circling particular takes, you should also circle the corresponding footage amounts to make it easier to add up the footage. These takes that are circled are called good (G) takes. The takes on the report that are remaining and have not been circled are called no good (NG). If, for some reason you circled a take and then the Director decides that she does not want it to be printed, draw slashes through the edges of the circle and write "Do Not Print" in the Remarks column. An example of a circled take that is not to be printed is shown on the camera report in Figure 3.10.

SCENE NO.	TAKE	DIAL	FEET	SD	REMARKS
54	1	70	70		
	2	120	50		
	3	160	40		
	4	210	50		DO NOT PRINT
54 A	1	250	40		
	2	280	30		

Figure 3.10 Marking a circled take to indicate that it is not to be printed.

At the bottom of most camera reports there are sections labeled G, NG, W, SE, and T. If the camera report does not have any of these sections on it, write them in yourself. Write the total footage for all circled takes on the report in the section marked G. Mark the total footage for all noncircled takes in the section marked NG. Then add up the totals for G and NG, and subtract this amount from the total amount of film loaded into the camera. This remaining amount of film, if any, would either be called *waste* or a *short end*. If it is waste, write it in the section marked W. If it is a short end, mark it in the section marked SE. A short end is a roll of film that is available for shooting that is left over from a full size roll. In other words, let's assume you loaded a 1000 foot roll of film into the camera and only shot 370 feet. This 370 feet is sent to the lab for developing with all other film shot during the day's shooting. The remaining 630 feet (1000 – 370) is left over and is called a *short end*. As a general rule for 35mm format, anything that is more than 100 feet is called a short end, and anything that is less than 100 feet is called waste. For 16mm format, anything more than 40 feet is a short end, and anything less than 40 feet is waste. The waste footage may either be thrown away or saved as a "dummy load" to use when scratch testing the magazines during the camera prep. The combined total of G, NG, W, and SE should equal the total amount of footage loaded in the magazine. This total amount would then be written in the section marked T.

Before removing the magazine from the camera, the 1st A.C. should cover the lens and run the camera for approximately 10 feet so that there will be a blank area of film at the end of the roll for safety reasons. If you remove the magazine immediately after the last take, you may fog the last few frames of the shot. This 10 feet of film can be included in the good or no good totals depending on what the last shot was. At the bottom of the report, after the last take, draw a diagonal line and write under the line the amount of footage that the roll ended at. For example, if the last dial reading on the camera report is 970 feet, the assistant will write "OUT AT 970." If the roll of film rolled out during the last take, write the amount of footage that the roll ended at or write "ROLLOUT." Whenever possible, I feel that it is better to reload the camera than to risk having a rollout, because when the film rolls out it is not good for the camera or the film. If you are in doubt as to whether you should reload or risk rolling out, check with the D.P. or Director and let one of them make the decision. For example, I have been in the situation when the previous take was 90 feet and there was 100 feet left on the camera. Because it is so close, I usually check with the D.P. or Director. If one of them chooses to try to do another take and the film rolls out, it is his or her responsibility. Whenever the film does roll out, write at the bottom of the report, "SAVE TAILS" as an indication to the lab to print the roll to the very end.

In addition, at the bottom of the report, write any developing instructions to the lab as given to you by the D.P. The instructions may include the following: develop normal, 1-Lite Print, time to gray scale, time to color chart, push one stop, print circle takes only, prep for video transfer, transfer circle takes only, and so on. Figures 3.5, 3.7, and 3.9 show the G, NG, W, SE, and T at the bottom of each camera report as well as the developing instructions.

Often the magazine may be loaded with a short end of film. The camera report should be marked in some way to indicate that it is not a full roll. The standard procedure for marking a camera report for a short end is to draw a diagonal line across the shooting part of the report to indicate it is a short end. This should be done before filling in the information on the report so that each time you look at the report, this diagonal line will remind you that it is a short end. The assistant will also write the footage in the lower left corner. A typical camera report for a short end is shown in Figure 3.11.

Each time you load a magazine with a fresh roll of film, a camera report should be attached to it. To save time, many 2nd A.C.s prepare a supply of camera reports with most of the heading information filled in ahead of time. Fill out as much information as possible in the heading so that the report is ready for shooting. This includes the production company, production title, Director, and Director of Photography.

PROD. COMPANY	Demi Monde Productions		
FILM TITLE	"Claire of the Moon"	DATE	10/29/91
CAMERAMAN	R. Sellars	DIRECTOR	N. Conn
CAMERA I.D. A	MAG NO. 10149	ROLL # 8	
☒ 35mm	☒ Color	☒ One Light	
☐ 16mm	☐ B & W	☐ Timed	
Type of Film/Emulsion	5298 - 237 - 4862	740'	
Processing - Normal ☒	Forced 1 Stop ☐	Forced 2 Stops ☐	

SCENE NO.	TAKE	DIAL	PRINT	REMARKS
17	(1)	130	(130)	
	2	220		
	(3)	290	(70)	
17 A	(1)	360	(70)	
	(2)	470	(110)	
42	1	530		
	(2)	580	(50)	
	3	630		
	(4)	680	(50)	
	(5)	720	(40)	
		OUT AT 720		
				G 520
				NG 200
	DEVELOP NORMAL			W 20
	1 - LIGHT PRINT			T 740
740				

Figure 3.11 Example of a completed camera report for a short end.

Some labs will preprint the heading information on the report for you. This saves time during the loading process. If you have some of the heading information filled in ahead of time, each time a magazine is loaded only a small amount of information needs to be added to the report. This will be discussed further in the section on loading magazines.

Sometimes it may be necessary to remove a partially shot roll of film from the camera, knowing that it will be used again later the same day. When doing this, remember not to break the film when removing the magazine. Attach the camera report to the magazine and place the magazine back in its case for later use. When the partially shot roll is placed back on the camera, the roll number remains the same. Be sure to inform the Script Supervisor that you are using a roll from previously in the day, and that it is not the next highest roll number but rather the same roll number as before.

Each time a new magazine is placed on the camera, the 2nd A.C. takes the camera report from the magazine and usually places it on the back of the slate. This gives a smooth writing surface to write out the report during shooting. Some assistants prefer to use a clipboard for the report. You may use whichever system is more convenient for you. Be sure to write clearly and legibly on the camera report so the people at the lab, and also the editors, will be able to read the report without any difficulty.

Magazines

A magazine is described as a lightproof container that is used to hold the film before and after exposure. There are two basic types of magazines in use today: coaxial and displacement. The coaxial magazine has two distinct compartments, one for the feed side and one for the take-up side. The magazine is called *coaxial* because the feed and take-up rolls share the same axis of rotation. Because there are two separate compartments, it is much easier to do the loading and unloading of the magazine. During the loading process, only the feed side needs to be loaded in the dark, and the take-up side can be loaded in the light. During the unloading process, the take-up side must be unloaded in the dark, and the feed side may be unloaded in the light, unless there is a short end. If there is any short end left in the magazine, then the feed side also must be unloaded in the dark during unloading.

A *displacement magazine* is so named because, as the film goes from the feed side to the take-up side, it is displaced from one side to the other. There are two different types of displacement magazines: the single-chamber displacement magazine and the double-chamber displacement magazine. On a displacement magazine the feed side is usually toward the front of the camera and the take-up side is toward the back of the camera when the magazine is in place. During shooting, as the film is displaced from the feed side to the take-up side, the film moves from the front of the camera to the back. This usually causes a shift in weight on the camera, so the camera must be periodically rebalanced.

The *double-chamber displacement magazine* may be handled the same as the coaxial magazine during the loading and unloading process. In other words, the feed side must be loaded in the dark and the take-up side loaded in the light during loading, and the take-up side unloaded in the dark during unloading. The *single-chamber displacement magazine* contains both the feed and the take-up sides of the magazine in the same compartment. Because of this, the entire loading and unloading process must be done in the dark. Most single-chamber displacement magazines are not able to hold a full roll of film on both the feed side and the take-up side at the same time. It is a good idea to be familiar with the loading and unloading procedure for as many different magazines as possible. Figure 3.12 shows two sides of a coaxial magazine. Figures 3.13 and 3.14 show the two different types of displacement magazines.

Loading Magazines

Before loading any magazine, clean it thoroughly to remove any dirt, dust, or film chips. Blow out the magazine using some type of compressed air. Be sure that the darkroom, changing bag, or changing tent is clean and that you have all the necessary items before you start to load any magazines. You should have camera tape, permanent felt tip markers, camera reports, extra cores, and so on. Most important, be sure that you have the correct film stock to load into the magazine.

Some magazines require a plastic core on the take-up spindle for the exposed film to be wound onto. You should have extra cores

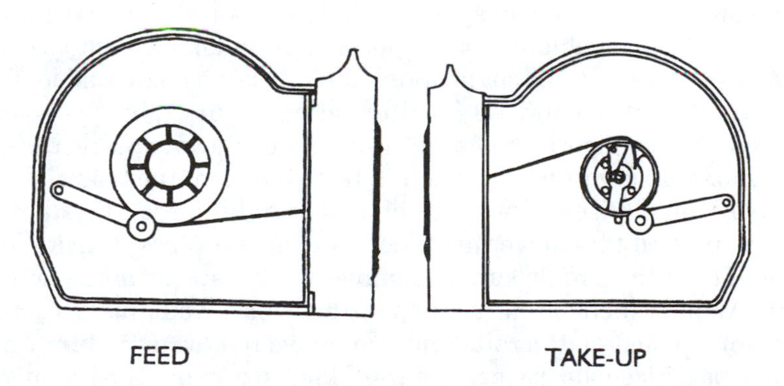

Figure 3.12 Arriflex 16SR coaxial magazine. (Courtesy of the Arriflex Corporation.)

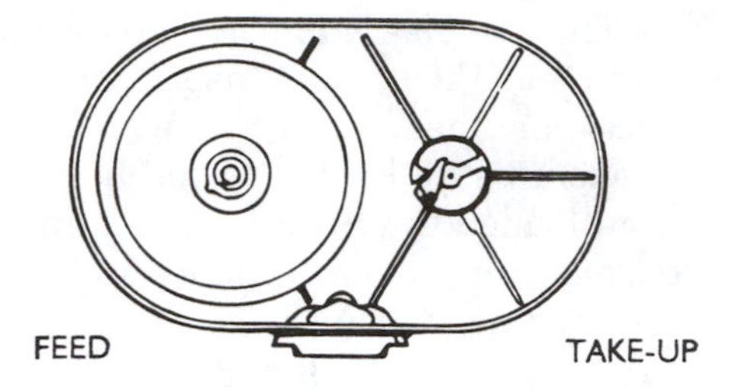

Figure 3.13
Single-chamber displacement magazine. (Courtesy of the Arriflex Corporation.)

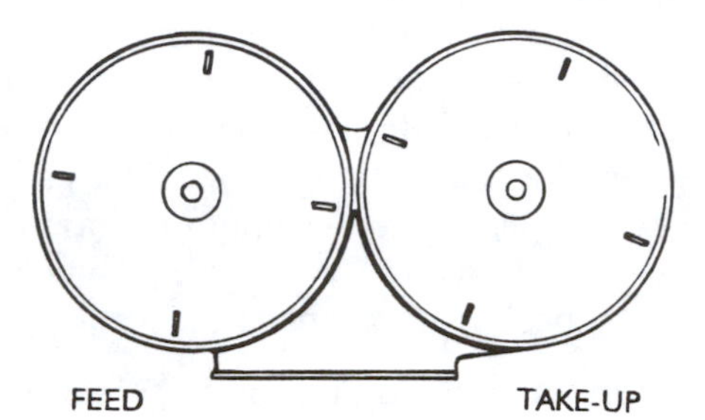

Figure 3.14
Double chamber displacement magazine.

available in this case. There are certain magazines that have a collapsible core on the take-up side. When the film is first placed on the collapsible core, it is inserted into a slot and locked in place. When you are ready to remove the exposed film from the take-up side, release the lock and the core collapses, which allows you to remove the roll of film easily from the magazine. Some cameras have the ability to accept internal loads without the use of a magazine. This is most often film that is wound onto a daylight spool. In this case, you should have extra daylight spools available for the exposed film to take up onto. Some magazines will accept a daylight spool, but it is not recommended. See Chapter 1, Figures 1.5 and 1.6 for illustrations of daylight spools and plastic cores. See Figure 3.15 for an illustration of a collapsible core.

When a fresh roll of raw stock is removed from the black bag, it will have a small piece of tape on the end to hold the roll together. It is very important to remove this piece of tape and place it inside the bottom of the film can. Be sure to remove all the tape from the end of the roll. Many camera or magazine jams have occurred due to a small amount of tape left on the roll. Once you have removed the film from the black bag, place the bag back in the film can and put the lid on the can. Do this to reduce the chance of the piece of tape or the black bag getting stuck in the magazine during the loading process.

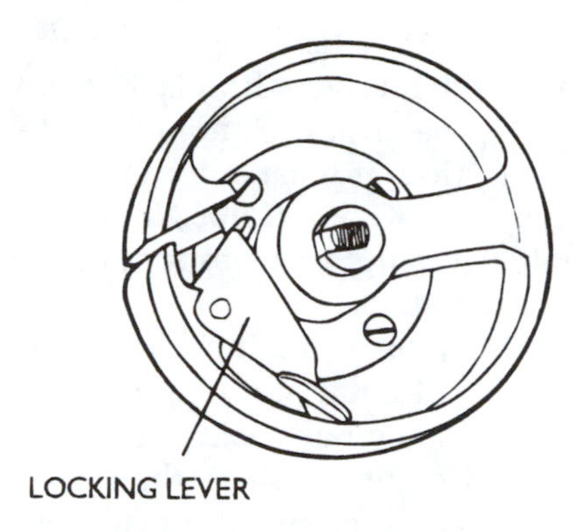

Figure 3.15
Collapsible core. (Reprinted from the *16SR Book*, with permission of the Arriflex Corporation.)

For ease of loading and threading the film, the end of it should have a straight edge, and it should be cut so that the cut bisects the perforations. Before loading a roll of film into a magazine, you may need to cut the film so that you bisect a perforation. This makes it easier to thread the film into a magazine, which has gears in it. Remember, you will need to do this in the dark so that you do not expose the film stock. Be very careful if you need to cut the film in a darkroom, and especially if you are using a changing bag or changing tent, so that you do not cut them. When threading the film onto a plastic core, fold the end of the film over onto itself, so that emulsion touches emulsion, approximately one inch from the end and insert it into the slot on the core. This gives the core a solid, thick piece of film to grab onto and keeps the film from slipping off the core.

Once you have loaded the magazine, an identification label must be placed on the lid to identify what is loaded in it. On a coaxial magazine, the identification label should be placed on the take-up side of the magazine. The identification label should contain the following information: production company, production title, date, footage, film type, emulsion number, roll number, and magazine number. If more than one person is loading magazines on the production, the initials of the loader also should be written on this piece of tape.

On many productions the identification label is usually made from a piece of one-inch white camera tape and a black permanent felt tip marker. Whenever possible many assistants will use a color coding system for labeling the magazines when they are using more than one type of film stock. For example, use white tape for slow-speed film, yellow tape for medium-speed film, and red tape for high-speed film.

This is especially useful when you are in a hurry because you don't have to take time to read the label to know what type of film is loaded in the magazine. The color of the tape will be an indication of what type of film is loaded. Table 3.2 is a suggestion of what color tape to use based on the currently available Eastman Kodak Color Negative and Fuji Color Negative films. This table is only a suggestion as to what color tape you should use for each film stock.

It is the system that I have used for many years and it has worked for me. You may adjust this to suit your particular shooting needs, depending on how many different film stocks you are using on your production. If you are unable to use a color coding system, then you should use one-inch white camera tape with a different color marking pen for each film stock. This will work just as well as the tape color coding system. The important thing to remember is that if you develop a system, stick with it and do not change it from production to production. The magazine label is usually six to eight inches in length and may look like the one shown in Figure 3.16.

Once the magazine has been loaded, place a piece of tape over the magazine lid as a safety measure. On many magazines you also need to have tape wrapped around the edges where the lid attaches to the magazine, to prevent light leaks and as a safety measure to keep the lid from coming off. This is especially important when filming outside in bright sunlight. If the magazine lid does not make a tight seal with the magazine, the very bright, direct sunlight can cause fogging on the film. If you are using the color code system for the magazine identification labels, the tape used for the lid will be the same color as that used for the identification label.

Table 3.2 Camera Tape Color Coding System When Using Various Films.

KODAK	ASA	FUJI	ASA	TAPE COLOR	INK COLOR
7245 / 5245	50 D	8621 / 8521	64 D	White	Black
7248 / 5248	100 T	8631 / 8531	125 T	White	Red
7287 / 5287	200 T	8651 / 8551	250 T	Yellow	Black
7293 / 5293	200 T			Yellow	Red
7297 / 7297	250 D	8661 / 8561	250 D	Blue	Black
7298 / 5298	500 T	8671 / 8571	500 T	Red	Black

D = Daylight T = Tungsten

A

Date	Production Company Production Title	Roll # Mag #
Footage	Film Type Emulsion Number	Loader Initials

B

10/29/91	Demi Monde Productions "Claire of the Moon"	Roll #7 Mag # 10146
1000'	5298 - 237 - 4862	DEE

Figure 3.16 (A) Information to be included on a magazine ID label. (B) Completed magazine ID label.

Once the magazine has been loaded and an identification label attached, attach a camera report to it. You should have filled in the heading portion previously, so now you only have to fill in the footage, film type, emulsion number, magazine number, and so on. Tape the camera report to the magazine so it is ready for use when the magazine is loaded onto the camera. When the magazine is then removed from its case for use, the camera report is already attached and you do not have to search to find a report. The report will be removed from the magazine and placed on the back of the slate for use during filming. Once you have finished using a particular magazine and roll of film, the camera report will then be reattached to it and the magazine placed back in the case. When you take it to the darkroom to unload and reload, the report is there so you can complete the unloading process without having to locate the report for that roll of film.

If the magazine is loaded with a short end, the footage amount on the identification label should be circled so that it stands out. In addition, you should make an additional, smaller identification label with only the footage marked on it, which is placed along side the larger identification label. When the magazine is loaded onto the camera, place this smaller piece of tape next to the footage counter of the camera. Each time you or the 1st A.C. look at the footage counter to obtain the dial readings, you will be reminded that there is a short end in the magazine. The short end identification label and smaller reminder label are shown in Figures 3.17 and 3.18.

After the magazines have been loaded, place them in their case, and attach another identification tape to the lid for each magazine

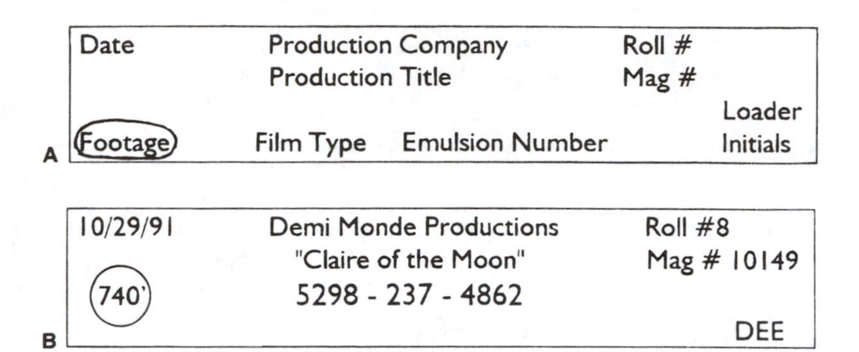

Figure 3.17 (A) Information to be included on a short-end ID label. (B) Completed short-end ID label.

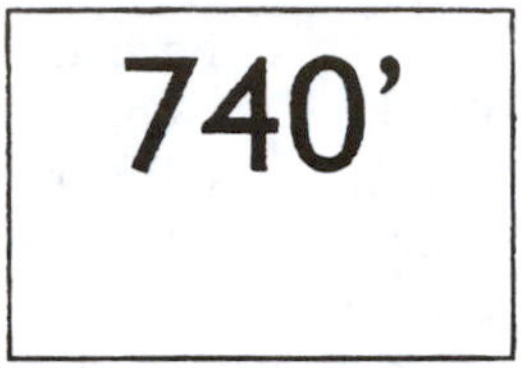

740'

Figure 3.18
Example of a short-end reminder label.

inside. This is just a small piece of tape with the footage amount written on it. If you are using the color coding system, use the same color tape used on the magazine identification label. By using the color coding system, you save time because you do not have to pick up the magazine or open the case to know what type of film is loaded. You are able to tell what type of film is in the magazine or case based on the color of tape used to make the labels.

During the day's shooting there will be many times when you will be required to unload and load magazines. When is the right time to go to the darkroom and reload any used magazines? It depends on the individual circumstances of the particular production that you are working on. In most cases, when there is a new lighting setup being done, the 2nd A.C. will usually have enough time to complete the reloading process. Always check with the 1st A.C. to see if it is all right to leave the set and do this job. The 1st A.C. usually has a lot on her mind, and may not realize that you have two or three magazines that need to be reloaded. Let her know the situation, and if it is convenient you will be allowed to reload. You should never wait until all of the magazines have been shot before reloading. This could result in the production having to stop shooting until you have time to load more film. If you keep on top of this throughout the day, the filmmaking process will go much smoother, and there should be no delays because a magazine is not ready. Try to find out ahead of time from the D.P. what film stock you will be using the next day and load the magazines before you go home for the night.

During the loading process, mistakes can happen, and there may be an instance when you accidentally expose a fresh roll of film to the light. In a rush you may open the magazine or the film can in the light, or possibly the lid of the magazine was not locked properly and unexpectedly opened in the light. You should immediately place this exposed roll of film in a black bag and put it back into the film can. Wrap the can with one-inch white camera tape and place a warning label on the can which reads, "FILM EXPOSED TO LIGHT—DO NOT USE." This warning should be written on the top of the can and also along the edge on the sealing tape. Place this can in a safe place away from all fresh raw stock, short ends, and exposed film that has been shot. You do not want to risk loading this film by accident and trying to shoot with it.

Finally, if this should ever happen to you, do not try to hide it. Notify the 1st A.C. immediately so that it can be brought to the attention of the D.P. and then to the production manager. By telling the appropriate people about this as soon as possible, you will show that you are a professional, and they should understand that it was probably only an accident and you did not do it intentionally. By trying to hide it you will only cause yourself problems, including losing your job and possibly not getting other jobs.

Unloading Magazines

Before unloading magazines check that you have everything needed to can out the film. You should have empty cans, black bags, black and white camera tape, and so on. Always remove the exposed film and place it in a black bag and can before removing any short end or waste. When unloading a roll of film that is on a plastic core, place the thumb of one hand on the inside edge of the core, and, using your other hand on the outside edge of the roll, gently lift the roll of film off the take-up spindle. As the roll starts to come up and off the spindle, slide your hand under the roll to keep the film from spooling off. When using a collapsible core, release the lock on the core, and using the same method as earlier, remove the roll of film from the take-up side of the magazine.

Always place the exposed film in a black bag and film can. Do not tape the end of the film to the roll. The standard rule for wrapping a can of exposed film is to use one-inch black camera tape. Some assistants use the red tape that is imprinted with the words "EXPOSED FILM—OPEN IN DARKROOM ONLY." Place the identification label that was on the magazine, on the film can, along with the top copy of the camera report. Be sure that the camera report is completely filled out with all the proper takes circled, and that the footage amounts are totaled for G, NG, W, SE, and T, and the lab instructions are written on it. Have the Script Supervisor double-check the report and initial it so that you are sure that the correct takes are circled. Once the can of exposed film is ready, keep it in a safe place away from any raw stock so that it does not get reloaded by mistake. See the section, Preparing Exposed Film for Delivery to the Lab, later in this chapter.

If there is any film left in the feed side of the magazine, remove it now. If it is a short end, it must be unloaded in the dark. Place it in a black bag and in a film can, and wrap it with tape. A general rule is to wrap all cans of unexposed film in one-inch white camera tape unless you are using the color coding system. But as I have mentioned before, if you have been using the color coding system, wrap the film can in the appropriate color tape. Label the can with the short end so that you know how much and what type of film is in the can. Using the appropriate color tape, place an identification label on the can with the following information: date, footage, film type, emulsion number, and the words "SHORT END." Put the initials of the assistant unloading the magazine on this label. In addition, write along the edge of the can, on the piece of sealing tape, the amount of footage in the can. The label for a can containing a short end looks like the one in Figure 3.19.

There may be times when you have to can up a roll of raw stock that was loaded but not used. When this happens, place the film in a black bag and into a film can. Seal the can with the appropriate color tape, and place an identification label on the can. This label should contain the following information: date, footage, film type, emulsion number, and the word "RECAN." The assistant's initials should also be placed on this label. Write the footage on the piece of sealing tape on the edge of the can. The label for a recan roll of film looks like the one in Figure 3.20.

Date SHORT END

Footage Film Type Emulsion Number Loader Initials

A

10/29/91 SHORT END

740' 5298 - 237 - 4862 DEE

B

Figure 3.19 (A) Information to be included on a short-end can label. (B) Completed short-end can label.

Date RECAN

Footage Film Type Emulsion Number Loader Initials

A

10/29/91 RECAN

1000' 5298 - 237 - 4862 DEE

B

Figure 3.20 (A) Information to be included on a recan label. (B) Completed label for a recan roll of film.

As when loading a magazine, accidents can also happen when unloading. If you should accidentally expose to light a roll that has been shot, you should tell the 1st A.C. and D.P. immediately. This situation is more serious than exposing a fresh roll of film. By exposing a roll that has already been shot, you are now requiring the production company to reshoot everything that was on that particular roll. This will most likely cost the production company a lot of unexpected money and may result in you losing your job, even if it was only an accident. The important thing to remember when loading, and especially when unloading, film is not to be rushed and to take your time. Rushing can only cause costly mistakes, not only to the production company but also to you if you lose the job. Don't let anybody rush you during the loading or unloading of any film magazine.

Using a Changing Bag or Changing Tent

If a darkroom is not available, you should have a changing bag or changing tent available for loading and unloading magazines. Most 2nd A.C.s have their own changing bag. If you don't, they are available for rental at most camera rental houses. Ask the production company to rent one along with the camera equipment. The changing bag is actually two bags, one within another. They are sewn together along the edges and along the sides of the two sleeves, which have elastic cuffs. At the top of each bag is a zipper so that you have access to the inside of the bag. With the zippers closed and your arms in the sleeves, you have a completely lightproof compartment for loading and unloading magazines.

The important thing to remember when using a changing bag is not to panic if something goes wrong. The area inside the bag is very small and confined, and you should take your time when working in the bag. One of the most common problems encountered when using a changing bag is that the core will come out of the center and the film will start to spool off the roll from the center. If this happens with the exposed roll of film, do not try to force the core back into the center of the roll. Carefully place the film back into the center of the roll without the core, and continue the unloading process normally. The lab does not need a core in the center of the roll to develop and process the film. If the core comes out of a roll of unexposed raw stock or a short end, do not force the core back into the center of the roll. Place this roll into a black bag and can, and start over with a new roll of film. If something does go wrong while you are working in a changing bag, remember, never open the bag until all film, whether exposed or unexposed, is in a black bag and in a film can.

Before using the bag, always turn it inside out and shake it to remove any loose film chips or material that may have become stuck in the bag. To check the bag for light leaks, place it over your head, and once your eyes have adjusted to the darkness, see if any light is leaking in. It is best to do this outside in bright sunlight so that you can better see any light leaking in. If any holes are found, they may be covered with black camera tape or gaffer tape, if they are not too large. When loading a magazine, place it in the inner bag with the can of unexposed raw stock. If necessary be sure to place an empty core on the take-up side of the magazine before placing it in the bag. Close both zippers along the top of the bag and then insert your arms into the elastic sleeves so that the elastic is past your elbows. When the magazine lid is removed, place it under the magazine to conserve space in the bag. Load the film in the usual manner and then place the lid back on the magazine, being careful not to catch the black bag between the magazine and the lid. Be sure that the lid is securely locked on the magazine before removing your arms from the bag and opening the zippers. Place the proper label on the magazine and tape the lid around the edges. Place the black bag back in the can so that it is ready when it is time to unload the magazine.

The unloading process is the reverse of the loading process, as described earlier. Place the magazine in the inner bag along with the appropriate amount of black bags and cans to can out any exposed film or any short ends. Again, remember to not remove your arms or open the bag until all film is placed in black bags and film cans.

Another new item that is used by many assistants is the film-changing tent. It is similar in size and shape to a changing bag, but instead of lying flat, it forms a lightproof tent to load and unload magazines. Figure 3.21 shows a changing bag and a changing tent.

When you are finished using the changing bag, always shake it out to remove any film chips or other foreign matter. Then close both zippers, lay the bag flat, and fold it according to the diagram in Figure 3.22. Folding the bag and storing it properly will help keep it clean and prevent it from becoming torn or ripped.

Whenever working as a 2nd A.C., you should never wear any type of clothing, such as loose sweaters, that could have fibers or

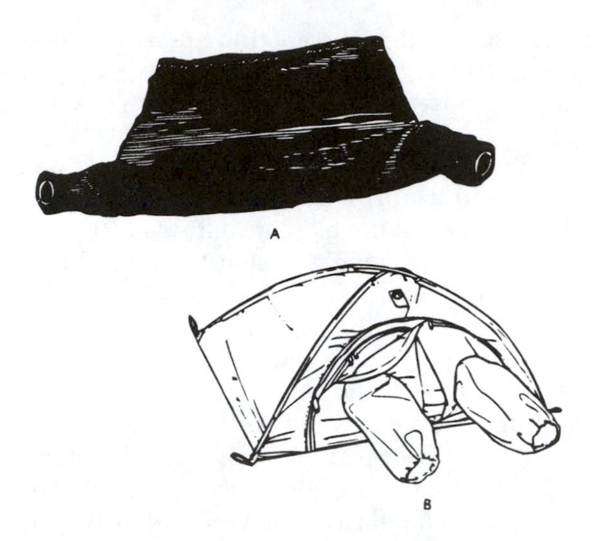

Figure 3.21 (A) Changing bag. (Reprinted from the *16SR Book*, with permission of the Arriflex Corporation.) (B) Changing tent. (Reprinted from the *Arri 35 Book*, with permission of the Arriflex Corporation.)

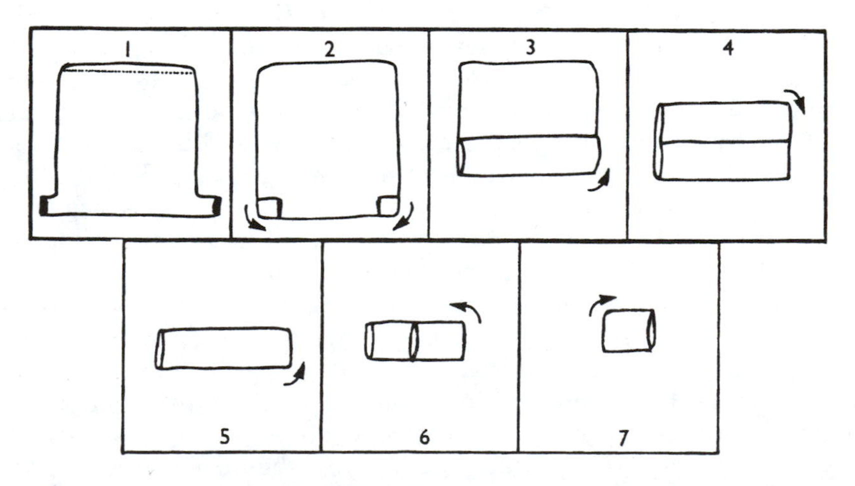

Figure 3.22 How to fold the changing bag.

threads that can get into the magazines. These small fibers or threads could scratch the film and create additional shooting time if scenes need to be reshot. This is especially important when working in a changing bag or changing tent. The process of placing your arms in the bag or tent could cause fibers or threads to become loose and fall into the magazine. In addition, if you wear a watch that has an illuminated dial, it should be removed before going into the darkroom or placing your hands in the changing bag or changing tent. The light from the dial could cause a slight fogging on the edges of the film. It is always better to take that extra step and be safe.

Setting Up the Camera

At the start of each shooting day the camera must be set up and made ready to shoot as quickly as possible. The actual setting up of the camera is handled by the 1st A.C. The 2nd A.C. stands nearby and hands pieces of equipment to the 1st A.C. as they are asked for. The procedure for setting up the camera is discussed in detail in the section, Setting Up the Camera, in Chapter 4.

Marking Actors

During rehearsals the 2nd A.C. places marks on the floor for each actor, for each position during the scene. These marks are used by the actors so that they know where to stand, by the 1st A.C. for focusing, and by the D.P. for lighting purposes. The marks are usually made with the colored paper tape that was included in the expendable list. If the floor or ground is seen in the shot, place tape marks for the rehearsal and then remove them or make them very small for the actual shot. If you are working outside or on a surface where you cannot place tape marks, use anything that is handy, such as leaves, sticks, twigs, rocks, and so on. When working on pavement or concrete, many assistants use a piece of chalk to make the marks for the actor. If more than one actor is in the scene, each actor's marks should be a different color. This makes it easier and less confusing for each actor. The most common type of mark used is the "T" mark, in the shape of the letter "T," and is usually three to five inches wide by three to five inches high. A "T" mark is placed with the top portion of the "T" just in front of the actor's toes and the center portion extending between the actor's feet. See Figure 3.23 for an example of an actor's "T" mark.

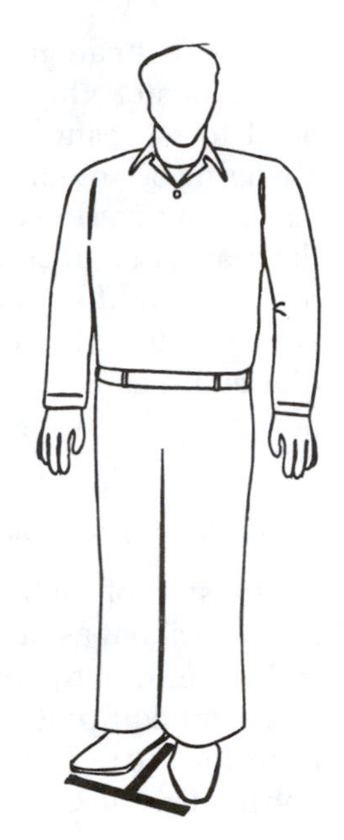

Figure 3.23
Example of a "T" mark.

Another type of mark is the toe mark. These are usually three- or four-inch-long pieces of tape placed at the end of each actor's foot. See Figure 3.24 for an example of toe marks.

A variation of the toe mark is an upside down "V" placed at each actor's foot. See Figure 3.25 for an example of an upside down "V" mark.

One other and more precise form of mark is a box placed completely around the actor's feet. See Figure 3.26 for an example of a box mark.

Slates

The slate is used to identify the pertinent information for each scene shot during the production. There are two basic types of slates: sync and insert. The *sync slate* is used anytime you are recording sound. The top part of the slate contains two pieces of wood painted with diagonal black and white lines. The top piece of wood is hinged to the bottom piece of wood, which is attached to the slate. These pieces of wood along with the slate are sometimes referred to as the *clapper*. An example of a sync slate is shown in Figure 3.27.

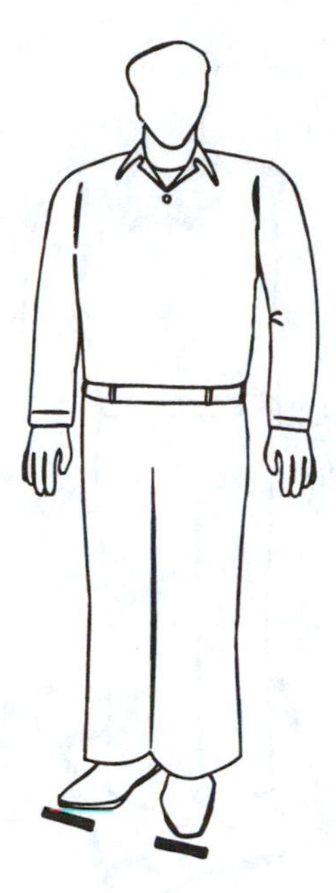

Figure 3.24
Example of toe marks.

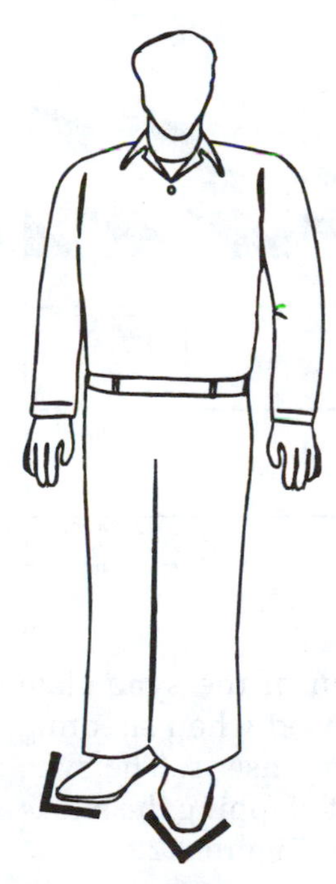

Figure 3.25
Example of an upside-down "V" mark.

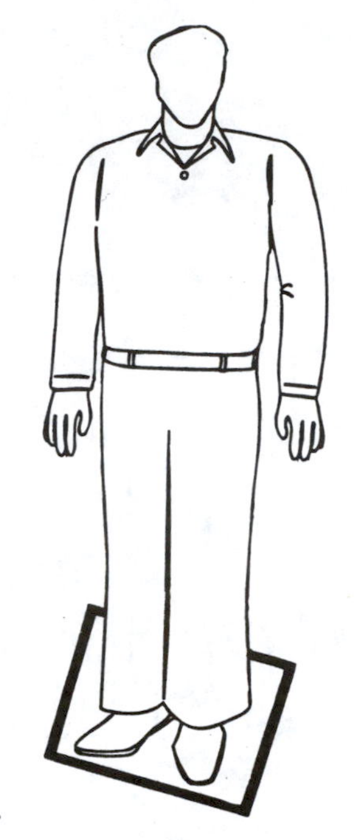

Figure 3.26
Example of a box mark.

Figure 3.27
Sync slate.

The *insert slate* is usually a smaller version of the sync slate without the wooden clapper attached. It is often used when shooting MOS shots or, as the name implies, when shooting inserts. The sync slate may be used for inserts or MOS shots without clapping the sticks together. An example of an insert slate is shown in Figure 3.28.

DATE	UNIT	CAMERA
SCENE	TAKE	ROLL
CAMERAMAN		PROD. NO.

Figure 3.28
Insert slate.

The information written on the sync and insert slate is usually as follows: production title, Director, Cameraman (D.P.), roll number, scene number, take number, int/ext, day/nite, and the date. When using the insert slate, the unit number and the production number may also be placed on the slate. The production title is the working title of the film during shooting. The Director is the name of the person who is directing the film. On the slate place the first initial and last name of the Director. Put the first initial and the last name of the D.P. on the slate next to camera. The roll number refers to the camera roll number that is being shot at the time. Get the scene and take numbers, which correspond to the scene in the script that is being filmed, from the Script Supervisor. "INT" means that you are filming on an interior set and "EXT" refers to an exterior set. "DAY/NITE" refers to the time of day that the scene takes place. The date is the month, day, and year that you are filming.

Before each shot, check with the Script Supervisor to find out what the scene number and take number is. Always write the numbers clearly on the slate to make it easier for the editor to read. When shooting a portion of a scene or a pick-up of action within a scene, a letter is usually added to the scene number. For example, if you are shooting scene number 15 and are only doing a small part of the scene, the scene number may be written as 15A, 15B, 15C, and so on. The Script Supervisor will tell you when to add a letter to the scene number and when to change scene numbers. Some letters that are not usually used for slating scene numbers are I, O, Q, S, and Z, which can resemble numbers when written hurriedly. The letter I resembles the number one (1), O and Q resemble the number zero (0), S resembles the number five (5), and Z resembles the number two (2). Check with the Script Supervisor to find out which letters they do not want you to use when slating scenes.

In Britain they do not use the scene numbers on the slate as is done in the United States. Instead, they are written as shot numbers, and the first shot on the first day of filming is shot number 1. The next is shot 2, then shot 3, and so on. They still use the take numbering system if a particular shot is done more than once.

When using more than one camera, the roll number is a combination of the camera letter and the number of the roll of film, for example, roll number A-1, A-2, A-3, B-1, B-2, and so on. If only one camera is used, the assistant may still use the A prefix for all roll numbers to avoid any confusion by the editors. If more than one camera is used, you should have a separate slate for each camera, and mark the lettering on each slate in a different color to distinguish one slate and camera from the other. For example, when using two cameras, the "A" camera slate may be labeled in red letters and the "B" camera slate may be labeled in blue letters.

Slating Procedures

During shooting, the 2nd A.C. is responsible for slating each shot, whether it is sound (sync) or silent (MOS). Remember to obtain the correct scene and take number from the Script Supervisor. The sound mixer also needs to know the scene and take number, and usually preslates the shot, which means that he starts the sound recorder and speaks into the microphone, calling out the scene and take number. When it is time to roll the shot, the recorder is ready to go.

The standard procedure for rolling the shot and slating a sound take is as follows. The Director or Assistant Director calls for quiet on the set and then for sound to roll. When the recorder is turned on and has reached the proper speed, the sound mixer calls out "SPEED." At this time the Camera Operator or 1st A.C. turns on the camera. When the camera reaches the proper speed, the operator or assistant calls out "SPEED." Now the 2nd A.C., who has been waiting patiently in front of the camera, calls out "MARKER," and claps the sticks together. While waiting for the camera to be turned on and to reach speed, the 2nd A.C. holds the slate in the shot with the clapper sticks held open at approximately a 45° angle to each other. After "MARKER" is called, the 2nd A.C. claps the sticks together.

It is the responsibility of the Camera Operator to frame the slate properly, but the 2nd A.C. should know where to place it so that the

Camera Operator does not have to move the camera in order to photograph the slate. Position the slate in such a way so that it is not too big or too small in the frame. A general rule for positioning the slate in front of the camera so that it can clearly be seen is as follows: For 35mm, hold the slate 1 foot from the camera for every 10 millimeters in focal length. For example, with a 50mm lens the slate should be held 5 feet away; for 35mm, 3½ feet; for 100mm, 10 feet; and so on. For 16mm, hold the slate 2 feet from the camera for every 10 millimeters in focal length. For example, with a 50mm lens the slate should be held 10 feet away. It is not necessary to measure this distance, only to approximate it so the slate fills up the frame. The slate should also be well lit so that the information on it can be read clearly. When filming in a dark set, use your small flashlight to illuminate the slate or possibly have an electrician set up a small light that is turned on for the slate and then turned off before the action of the scene begins. The 1st A.C. adjusts focus for the slate so that it is easy to read and not blurry and out of focus. Once the slate has been photographed, the focus will be shifted back to the correct place for the scene.

When clapping the sticks together, remember to hold the slate perfectly still. Many assistants who are new at slating will move the slate in a downward motion when clapping the sticks. This causes a blurred image, making it difficult for the editor to read the slate. Another good practice to follow is to never cross the frame after slating, if it is possible. If you slate from the right, then exit to the right, and if you slate from the left then exit to the left. This is a courtesy to the actors as well as the Camera Operator. Sometimes it may not be possible to do this because of lights, C-stands, or set walls or furnishings. Be sure to watch where you go after a shot. Many times a shot is ruined because a 2nd A.C. does not watch where he moves after slating the shot, and ends up standing in front of a light, causing a shadow on the actor, or moves in the way of the dolly.

When slating a close-up shot of the actor, it is usually necessary to hold the slate very close to the actor's face. Don't clap the sticks so loudly that you startle the actor. The sound microphones are very sensitive, so a light clap is sufficient. Often the slate will not be framed properly, or it may be missed completely by the Camera Operator, and he will call for second sticks or second marker. When this happens, insert the slate quickly into the shot, and when the Camera Operator tells you that it is framed properly, call out "SECOND STICKS" or "SECOND MARKER" before clapping the sticks together. Whenever you do second sticks, be sure to note it in the Remarks column of the camera report.

There are also situations in which it is not possible or practical to clap the slate at the beginning of the scene. When this happens you do what is called a *tail slate*. The tail slate is clapped the same way as a head slate, the only difference is that the slate is held upside down in the frame. If you know before the shot that you will be doing a tail slate, photograph the slate before the shot just for identification of the roll, scene, and take numbers. Tell the sound mixer whenever you are doing a tail slate. The sound and camera will roll normally, but when the director calls "cut," the camera and sound recorder are kept running. The 2nd A.C. calls out "TAIL SLATE" before clapping the sticks together. Always make note of a tail slate in the Remarks column of the camera report.

If you use two cameras on a production and they both will be rolling together, there are two ways that you may slate the scene. You may slate by doing what is called a *common slate* or you may do what is called *separate slates*. When doing separate slates, each camera is slated individually, using the correct slate for each camera. When sound and cameras are rolling, the cameras are slated in order. Each slate is held in front of its respective camera. The 2nd A.C. slates the A camera first, then the B camera, then the C camera, and so on. When doing separate slates the 2nd A.C. calls out the camera letter before clapping the sticks. For example, when using two cameras labeled "A" and "B," the 2nd A.C. calls "A camera marker" before slating the A camera, and then "B camera marker" before slating the B camera. When doing a common slate, photograph an identification slate before the shot, showing the correct roll, scene, and take numbers for each camera. When sound and cameras are rolling, only one slate is used, and it is held so that the back of the slate is facing both cameras. The 2nd A.C. calls out "A and B cameras, common marker," before clapping the sticks together. Many 2nd A.C.s have a large set of only the stick part of the slate. These larger sticks are easier to see and tell the editor that it is a common slate for more than one camera.

There are a number of ways that you may slate an MOS shot. Since there is no sound for an MOS shot, you want to be sure that the editor knows that the sticks have not been clapped. The most obvious way to do this is to hold the slate with the sticks closed and your hand over them. Many assistants hold the sticks in an open position with their hand in between the two sticks to indicate that they have not or cannot be clapped together. In any case, when slating an MOS shot, be sure to indicate it clearly on the slate and also on the camera report.

Changing Lenses, Filters, and Magazines

Change or add any piece of equipment on the camera as quickly as possible. The usual procedure for changing anything on the camera is as follows. When the D.P. or Camera Operator requests a piece of equipment, the 1st A.C. tells the 2nd A.C. While the 2nd A.C. obtains the new item from the case, the 1st A.C. removes the old item from the camera and prepares the camera for the new one. When the 2nd A.C. brings the new item to the camera, it is exchanged for the old item with the 1st A.C. While the 1st A.C. places the new item on the camera, the 2nd A.C. places the old item back in the case. Whenever the 1st A.C. calls out a piece of equipment to you, it should always be repeated back so that she is sure that you heard it and heard it correctly.

Before handing a new lens to the 1st A.C. set the aperture to its widest opening. Whenever handing off pieces of equipment to each other, it is a good idea to call out "got it" or some other signal as an indication to the other assistant that it is all right to let go of the item. This is especially important when exchanging lenses. Many times lenses or filters are dropped and damaged because one assistant released her grip on the item before the other assistant had a firm hold on it.

Also, remember never to leave an equipment case opened when you are away from it. If a case is in use lock at least one of the latches. This makes it easier to open when you have to go back into the case. Any case that is not in use should have both latches secured. This is a good safety habit to get into because if you leave the case unlatched and someone tries to pick it up and move it while you are away from it, the contents could spill out and become damaged. If this did happen, it would be blamed on the person who left the case unlatched and not the person who tried to pick it up and move it.

Always check lenses and filters for scratches and dirt or dust before handing them to the 1st A.C. Tell the 1st A.C. that the lens or filter needs to be cleaned when handing it to her. Once the D.P. or Camera Operator has approved the new item, it then may be removed and cleaned by either assistant. When changing from a prime lens to a zoom lens or from a zoom lens to a prime lens, you should bring both lens cases to the camera to make the change quicker and easier. Once the change has been completed, you may then return both cases to the cart or storage area. Also, when changing lenses you may have to change the lens support rods and support brackets, because of the physical size or weight of the lens. When bringing the lens from the case, the 2nd A.C. should remember to bring the appropriate lens support rods and support brackets when required.

When changing magazines write the new roll number on the identification label, remove the camera report from the magazine, and place it on the back of the slate. If the magazine contains a short end, remind the 1st A.C. of this and tell her to place the small reminder tape next to the footage counter.

Using a Video Tap and Monitor

Today most productions are using a video tap incorporated into the film camera so that the Director can view the shot on a monitor while it is being filmed. During the camera prep all of the needed accessories and cables should have been obtained for the video system. During each shooting day, the camera will be moved to many different locations and sets for the various shots. Whenever the camera is to be moved, the 1st A.C. should disconnect the video cable from the monitor to the camera. It is the responsibility of the 2nd A.C. to be sure that the monitor is moved along with the camera, set up, and connected for each shot. On larger productions, a separate person, such as a Production Assistant or possibly a Camera Trainee may be responsible for moving and setting up the monitor for each shot.

Preparing Exposed Film for Delivery to the Lab

At the end of each shooting day, all the film shot must be sent to the lab for processing. As I mentioned in the section on unloading magazines, all exposed cans of film should have the proper identification label on them, along with the top copy of the camera report. This assists the lab so that they know which shots to print, and if there are any special instructions for them to follow during the developing process. Check with the Script Supervisor regarding the circled or printed takes. The best time to check with the Script Supervisor is at the time you place a new magazine on the camera. When you take the old magazine off the camera, give the camera report for that roll to the Script Supervisor for checking. She will check it to be sure that the correct takes are circled, and return it to you. Write the G, NG, W, SE, and T on the camera report.

You should also place an additional piece of tape on the can, with the developing instructions to the lab printed on it. Some examples of specific developing instructions include: "DEVELOP NOR-

MAL—PREP FOR TELECINE," "DEVELOP NORMAL—ONE LITE PRINT," "PUSH ONE STOP," "DEVELOP ONLY—NO WORK PRINT." There are many other types of developing instructions that may be used. Be sure to check with the D.P. before sending any film to the lab. In place of the magazine identification label and developing instructions label, some assistants may use a preprinted label that is filled in with the appropriate information and placed on the exposed film can. This label may look like the one shown in Figure 3.29.

Send the exposed film to the lab as soon as possible. Each lab usually has a specific cutoff time each day for when the film must be delivered in order for it to be ready the following day. As a 2nd A.C. you should know the cutoff times for the lab you are using. Until it is ready to be sent, keep the exposed film in a cool, dry place away from any direct sunlight and away from any raw stock so that it does not get loaded by mistake. See the section in this chapter, Ordering Additional Film Stock, for information on the proper care and storage of film stock.

Once you are ready to send the film, stack the cans four or five high, and tape them together. You should invert the top can so that you do not tape over the attached camera report. If the film is to be shipped, place it in a sturdy cardboard box, and fill any unused space with crumpled newspaper to prevent the cans from moving around

Date ____________________

Production Company ____________________

Production Title ____________________

Footage __________ Film Type ____________________

Camera ____________ Mag # __________ Roll # __________

DEVELOPING INSTRUCTIONS

☐ Develop Normal ☐ Prep For Telecine

☐ One Light Work Print ☐ Other ____________________

Figure 3.29 Example of a blank film can label.

during shipping. If film is to be shipped, label the box on all sides "EXPOSED FILM—KEEP FROM RADIATION" or "EXPOSED FILM—DO NOT X-RAY."

Be especially careful when transporting film on a plane. Although the x-ray equipment used to check baggage emits a very low-level dose of radiation, it can still cause a fogging on the film. Whenever you are going to transport film on a plane, I recommend carrying it on by hand and requesting that it be inspected by hand. Be sure to have your changing bag or changing tent available as the security officer may want to open some of the cans to be sure that it is indeed film inside of them.

Film Inventory and Record of Film Shot

Throughout the production you will be receiving shipments of film stock. You should have a supply of daily film inventory sheets so that you may keep an inventory of all film stock received. In most cases the production company needs an inventory of each different film stock as well as a grand total for all film stocks combined. For example, if you are using Eastman Kodak Color Negative 5245 and 5298 on your production, you would have three separate totals for the film inventory, one for 5245, one for 5298, and one for the combined total of both. When keeping the inventory, you may use a standard inventory form or make up one of your own. It is very important to keep accurate records in case there are any questions during the production. Some examples of different types of daily film inventory forms can be found in Figures 3.30–3.32.

At the end of each shooting day, after the equipment has been packed up and the film sent to the lab, the 2nd A.C. prepares a daily film inventory form that contains the following information: film received; each roll number shot; a breakdown of G, NG, W, SE, and T for each roll; film on hand at the end of the day; totals for each day; and a running total for the entire production. Be very careful when totaling up these numbers, because it is very important to the production office to account for every foot of film used on the production.

Once these reports have been filled out, give a copy to the production office along with copies of the camera reports for each roll. You should also keep a copy of any reports for the camera department in case there are any questions later. When using more than one camera, keep separate totals for each camera as well as combined totals for all cameras.

PRODUCTION ______________________ DATE __________

FILM STOCK __________ DAY # __________

ROLL #	GOOD	NO GOOD	WASTE	TOTAL(1)	SE	TOTAL(2)

TOTALS	GOOD	NO GOOD	WASTE	TOTAL(1)	FILM ON HAND
TODAY					PREVIOUS
PREVIOUS					TODAY(+)
TO DATE					TODAY(-)
					TO DATE

FILM STOCK __________

ROLL #	GOOD	NO GOOD	WASTE	TOTAL(1)	SE	TOTAL(2)

TOTALS	GOOD	NO GOOD	WASTE	TOTAL(1)	FILM ON HAND
TODAY					PREVIOUS
PREVIOUS					TODAY(+)
TO DATE					TODAY(-)
					TO DATE

TOTAL FILM USE - ALL FILM STOCKS

TOTALS	GOOD	NO GOOD	WASTE	TOTAL(1)	TOTAL FILM ON HAND
TODAY					PREVIOUS
PREVIOUS					TODAY(+)
TO DATE					TODAY(-)
					TO DATE

Figure 3.30 Daily film inventory form #1.

PRODUCTION ______________________________ DATE ________________

DAY # __________

FILM TYPE	LOAD TYPE	ROLL #	GOOD	NO GOOD	WASTE	TOTAL	SE
			GOOD	NO GOOD	WASTE	TOTAL	SE
		PREVIOUS					
		TODAY (+)					
		TO DATE					

UN-EXPOSED FILM ON HAND											
	FILM TYPE				FILM TYPE						TOTAL
	1000' LOADS	400' LOADS	SE		1000' LOADS	400' LOADS	SE		ON HAND	ON HAND	ON HAND
PREVIOUS											
TODAY (+)											
TODAY (-)											
TO DATE											

Figure 3.31 Daily film inventory form #2.

PRODUCTION ______________________ DATE ____________

DAY # ____________

NEG.	ROLL #	AMOUNT	GOOD	NO GOOD	WASTE	TOTAL	SE
TOTALS							

	( )	( )	( )	TOTAL
BALANCE (1)				
(+) REC'D TODAY				
BALANCE (2)				
(-) USED TODAY				
BALANCE (3)				

Figure 3.32 Daily film inventory form #3.

Completing Film Inventory Forms

The following is an example to show you how to fill out the daily film inventory forms and how each day's totals relate to the next day's daily film inventory form.

Example: You have been hired as the 2nd Assistant Cameraman on a feature film. The film is called "Claire of the Moon" and is being produced by Demi Monde Productions. The Director is Nicole Conn and the Director of Photography is Randy Sellars. The D.P. has decided to use two film stocks for this shoot, Eastman Kodak 5245 and 5298. He will be doing some handheld shots, so he will need 400 foot rolls in addition to 1000 foot rolls.

On the first day of production the following film stock is received:

Eastman Kodak 5245-148-0739	5200 feet	4–1000 foot rolls, 3–400 foot rolls
Eastman Kodak 5298-237-4862	5400 feet	3–1000 foot rolls, 6–400 foot rolls

On the second day of production, the following film stock is received:

Eastman Kodak 5245-148-0739	7000 feet	5–1000 foot rolls, 5–400 foot rolls
Eastman Kodak 5298-237-4862	5000 feet	3–1000 foot rolls, 5–400 foot rolls

Figures 3.33–3.45 show the completed camera reports and completed daily film inventory forms for day one and day two. So that you may become familiar with the different styles of camera reports, this example uses each of the styles for each day. On an actual production you would only use one camera report style from a single lab and not mix them.

Company	Demi Monde Productions		
Pic. Title	"Claire of the Moon"		
Director	N. Conn		
Cameraman	R. Sellars		
Mag. No.	10161	Roll No.	1
Footage	400'	Date Exposed	10/28/91
Film Type	5245	Emulsion No.	148 - 0739
Develop	☐ Normal	☐ Other	

SCENE NO.	TAKE	DIAL	FEET	SD	REMARKS	SCENE NO.	TAKE	DIAL	FEET	SD	REMARKS
32	1	80	80								
	(2)	120	(40)								
	(3)	160	(40)			DEVELOP NORMAL					
32 A	1	210	50			1 - LITE PRINT					
	(2)	270	(60)								
32 B	(1)	300	(30)								
32 C	1	320	20								
	(2)	360	(40)								
	OUT AT 360'										G 210
											NG 150
											W 40
											T 400

Figure 3.33 Completed camera report for roll #1.

PROD. COMPANY Demi Monde Productions		
FILM TITLE "Claire of the Moon"		DATE 10/28/91
CAMERAMAN R. Sellars		DIRECTOR N. Conn
CAMERA I.D. A	MAG NO. 10109	ROLL # 2
☐ 35mm	☐ Color	☐ One Light
☐ 16mm	☐ B & W	☐ Timed
Type of Film/Emulsion 5245 - 148 - 0739		400'
Processing - Normal ☐	Forced 1 Stop ☐	Forced 2 Stops ☐

SCENE NO.	TAKE	DIAL	PRINT	REMARKS
32 C	(1)	40	(40)	
	(4)	80	(40)	
32 D	(1)	150	(70)	
36	1	170		
	(2)	190	(20)	
	3	200		
36 A	1	220		
	(2)	240	(20)	
	3	260		
	(4)	280	(20)	
36 B	(1)	350	(70)	
	2	390		
	OUT AT 390'			FOOTAGE
				GOOD 280
DEVELOP NORMAL				N.G. 110
1 - LITE PRINT				WASTE 10
				TOTAL 400

Figure 3.34 Completed camera report for roll #2.

DATE 10/28/91	
COMPANY Demi Monde Productions	
DIRECTOR N. Conn	CAMERAMAN R. Sellars
PRODUCTION TITLE "Claire of the Moon"	
MAGAZINE NO. 10146	ROLL NO 3

TYPE OF FILM AND TYPE OF DAILIES
PLEASE CIRCLE

35MM COLOR	35MM B/W	16MM COLOR	16MM B/W
ONE LIGHT	ONE LIGHT	ONE LIGHT	ONE LIGHT
TIMED DAILIES	TIMED DAILIES	TIMED DAILIES	TIMED DAILIES

TYPE OF FILM/EMULSION 5298 - 237 - 4862 1000'

PRINT CIRCLE TAKES ONLY:

	TAKES				
SCENE NO.	1^5	2^6	3^7	4^8	REMARKS
79	30	(40)	(40)	30	30 70
79 A	(20)				110 140
79 B	0	(30)			160 160
79 C	20	(20)			190 210
79 D	30	(30)			230 260
79 E	(10)	(10)			290 300
79 F	(20)				310 330
94	(20)	20	(40)		350 370
99	60	(80)	50	(70)	410 470
					550 600
					670
	OUT AT 670'				
					G 430
DEVELOP NORMAL					NG 240
1 - LITE PRINT					W 0
					T 670
					SE 330

Figure 3.35 Completed camera report for roll #3.

PRODUCTION "Claire of the Moon" DATE 10/28/91

FILM STOCK 5245 DAY # 1

ROLL #	GOOD	NO GOOD	WASTE	TOTAL(1)	SE	TOTAL(2)
1	210	150	40	400	0	400
2	280	110	10	400	0	400

TOTALS	GOOD	NO GOOD	WASTE	TOTAL(1)	FILM ON HAND	
TODAY	490	260	50	800	PREVIOUS	0
PREVIOUS	0	0	0	0	TODAY(+)	5,200
TO DATE	490	260	50	800	TODAY(-)	800
					TO DATE	4,400

FILM STOCK 5298

ROLL #	GOOD	NO GOOD	WASTE	TOTAL(1)	SE	TOTAL(2)
3	430	240	0	670	330	1000

TOTALS	GOOD	NO GOOD	WASTE	TOTAL(1)	FILM ON HAND	
TODAY	430	240	0	670	PREVIOUS	0
PREVIOUS	0	0	0	0	TODAY(+)	5,400
TO DATE	430	240	0	670	TODAY(-)	670
					TO DATE	4,730

TOTAL FILM USE - ALL FILM STOCKS

TOTALS	GOOD	NO GOOD	WASTE	TOTAL(1)	TOTAL FILM ON HAND	
TODAY	920	500	50	1,470	PREVIOUS	0
PREVIOUS	0	0	0	0	TODAY(+)	10,600
TO DATE	920	500	50	1,470	TODAY(-)	1,470
					TO DATE	9.130

Figure 3.36 Completed daily film inventory form #1 for day #1.

PRODUCTION "Claire of the Moon"

DATE 10/28/91

DAY # 1

FILM TYPE	LOAD TYPE	ROLL #	GOOD	NO GOOD	WASTE	TOTAL	SE
5245	400	1	210	150	40	400	0
5245	400	2	280	110	10	400	0
5298	1000	3	430	240	0	670	330
			GOOD	NO GOOD	WASTE	TOTAL	SE
		PREVIOUS	0	0	0	0	0
		TODAY (+)	920	500	50	1,470	330
		TO DATE	920	500	50	1,470	330

UN-EXPOSED FILM ON HAND											
	FILM TYPE 5245				FILM TYPE 5298				5245	5298	TOTAL
	1000' LOADS	400' LOADS	SE		1000' LOADS	400' LOADS	SE		ON HAND	ON HAND	ON HAND
PREVIOUS	0	0	0		0	0	0		0	0	0
TODAY (+)	4,000	1,200	0		3,000	2,400	330		5,200	5,730	10,930
TODAY (-)	0	800	0		1,000	0	0		800	1,000	1,800
TO DATE	4,000	400	0		2,000	2,400	330		4,400	4,730	9,130

Figure 3.37 Completed daily film inventory form #2 for day #1.

PRODUCTION "Claire of the Moon" DATE 10/28/91

DAY # 1

NEG.	ROLL #	AMOUNT	GOOD	NO GOOD	WASTE	TOTAL	SE
5245	1	400	210	150	40	400	0
5245	2	400	280	110	10	400	0
5298	3	1000	430	240	0	670	330
TOTALS		1,800	920	500	50	1,470	330

	(5245)	(5298)	( )	TOTAL
BALANCE (1)	0	0		0
(+) REC'D TODAY	5,200	5,400		10,600
BALANCE (2)	5,200	5,400		10,600
(-) USED TODAY	800	670		1,470
BALANCE (3)	4,400	4,730		9,130

Figure 3.38 Completed daily film inventory form #3 for day #1.

Company	Demi Monde Productions		
Pic. Title	"Claire of the Moon"		
Director	N. Conn		
Cameraman	R. Sellars		
Mag. No.	10149	Roll No.	4
Footage	1000'	Date Exposed	10/29/91
Film Type	5245	Emulsion No.	148 - 0739
Develop	☐ Normal	☐ Other	

SCENE NO.	TAKE	DIAL	FEET	SD	REMARKS	SCENE NO.	TAKE	DIAL	FEET	SD	REMARKS
24	1	70	70				3	670	40		
	(2)	160	(90)			97	(1)	690	(20)		
	3	200	40				2	720	30		
	(4)	240	(40)			135	(1)	780	(60)		
24 A	(1)	290	(50)				(2)	830	(50)		
	2	340	50				3	870	40		
	(3)	400	(60)			146	1	910	40		
10	1	430	30				(2)	940	(30)		
	2	450	20								
	3	470	20								
	(4)	500	(30)				OUT AT 940'				G 490
	5	530	30								NG 450
	(6)	560	(30)								W 60
10 A	1	600	40			DEVELOP NORMAL					T 1000
	(2)	630	(30)			1 - LITE PRINT					

Figure 3.39 Completed camera report for roll #4.

DATE 10/29/91
COMPANY Demi Monde Productions
DIRECTOR N. Conn CAMERAMAN R. Sellars
PRODUCTION TITLE "Claire of the Moon"
MAGAZINE NO. 10161 ROLL NO 5
TYPE OF FILM AND TYPE OF DAILIES PLEASE CIRCLE
35MM COLOR 35MM B/W 16MM COLOR 16MM B/W
ONE LIGHT ONE LIGHT ONE LIGHT ONE LIGHT
TIMED DAILIES TIMED DAILIES TIMED DAILIES TIMED DAILIES
TYPE OF FILM/EMULSION 5245 - 148 - 0739 400'
PRINT CIRCLE TAKES ONLY:

	TAKES				
SCENE NO.	1 [5]	2 [6]	3 [7]	4 [8]	REMARKS
7	(60)	20	(70)	40	60 80
7 A	40	(30)	(30)	10	150 190
	20				230 260
7 B	(30)	(20)	20		290 300
					320 350
					370 390
	OUT AT 390'				
					G 240
DEVELOP NORMAL					NG 150
1 - LITE PRINT					W 10
					T 400

Figure 3.40
Completed camera report for roll #5.

Company	Demi Monde Productions		
Pic. Title	"Claire of the Moon"		
Director	N. Conn		
Cameraman	R. Sellars		
Mag. No.	10014	Roll No.	6
Footage	1000'	Date Exposed	10/29/91
Film Type	5298	Emulsion No.	237 - 4862
Develop	☐ Normal	☐ Other	

SCENE NO.	TAKE	DIAL	FEET	SD	REMARKS	SCENE NO.	TAKE	DIAL	FEET	SD	REMARKS
5	1	20	20			57	(2)	810	(40)		
	2	40	20				3	850	40		
	(3)	130	(90)				(4)	920	(70)		
	(4)	200	(70)								
	(5)	280	(80)				OUT AT 920'				
5 A	1	300	20								
	(2)	400	(100)								
5 B	(1)	440	(40)			DEVELOP NORMAL					
	2	520	80			1 - LITE PRINT					
	(3)	550	(30)								
12	1	590	40								G 680
	(2)	660	(70)								NG 240
	(3)	730	(70)								W 80
	4	750	20								T 1000
57	(1)	770	(20)								

Figure 3.41 Completed camera report for roll #6.

PROD. COMPANY Demi Monde Productions		
FILM TITLE "Claire of the Moon"		DATE 10/29/91
CAMERAMAN R. Sellars		DIRECTOR N. Conn
CAMERA I.D. A	MAG NO. 10250	ROLL # 7
☐ 35mm	☐ Color	☐ One Light
☐ 16mm	☐ B & W	☐ Timed
Type of Film/Emulsion 5298 - 237 - 4862 1000'		
Processing - Normal ☐	Forced 1 Stop ☐	Forced 2 Stops ☐

SCENE NO.	TAKE	DIAL	PRINT	REMARKS
33	1	30		
	2	50		
	(3)	70	(20)	
	(4)	100	(30)	
33 A	1	130		
	(2)	170	(40)	
33 B	(1)	200	(30)	
	2	220		
	3	240		
	(4)	270	(30)	
	5	290		
	(6)	320	(30)	
33 C	1	360		
	(2)	420	(60)	
107	(1)	490	(70)	
	2	520		
	(3)	580	(60)	
	OUT AT 580'			FOOTAGE
				GOOD 370
DEVELOP NORMAL				N.G. 210
1 - LITE PRINT				WASTE 0
				TOTAL 580
				SE 420

Figure 3.42 Completed camera report for roll #7.

PRODUCTION "Claire of the Moon" DATE 10/29/91

FILM STOCK 5245 DAY # 2

ROLL #	GOOD	NO GOOD	WASTE	TOTAL(1)	SE	TOTAL(2)
4	490	450	60	1000	0	1000
5	240	150	10	400	0	400

TOTALS	GOOD	NO GOOD	WASTE	TOTAL(1)	FILM ON HAND	
TODAY	730	600	70	1,400	PREVIOUS	4,400
PREVIOUS	490	260	50	800	TODAY(+)	7,000
TO DATE	1,220	860	120	2,200	TODAY(-)	1,400
					TO DATE	10,000

FILM STOCK 5298

ROLL #	GOOD	NO GOOD	WASTE	TOTAL(1)	SE	TOTAL(2)
6	680	240	80	1000	0	1000
7	370	210	0	580	420	1000

TOTALS	GOOD	NO GOOD	WASTE	TOTAL(1)	FILM ON HAND	
TODAY	1,050	450	80	1,580	PREVIOUS	4,730
PREVIOUS	430	240	0	670	TODAY(+)	5,000
TO DATE	1,480	690	80	2,250	TODAY(-)	1,580
					TO DATE	8,150

TOTAL FILM USE - ALL FILM STOCKS

TOTALS	GOOD	NO GOOD	WASTE	TOTAL(1)	TOTAL FILM ON HAND	
TODAY	1,780	1,050	150	2,980	PREVIOUS	9,130
PREVIOUS	920	500	50	1,470	TODAY(+)	12,000
TO DATE	2,700	1550	200	4,450	TODAY(-)	2,980
					TO DATE	18,150

Figure 3.43 Completed daily film inventory form #1 for day #2.

PRODUCTION "Claire of the Moon" DATE 10/29/91

DAY # 2

FILM TYPE	LOAD TYPE	ROLL #	GOOD	NO GOOD	WASTE	TOTAL	SE
5245	1000	4	490	450	60	1000	0
5245	400	5	240	150	10	400	0
5298	1000	6	680	240	80	1000	0
5298	1000	7	370	210	0	580	420
			GOOD	NO GOOD	WASTE	TOTAL	SE
		PREVIOUS	920	500	50	1,470	330
		TODAY (+)	1,780	1,050	150	2,980	420
		TO DATE	2,700	1550	200	4,450	750

UN-EXPOSED FILM ON HAND											
	FILM TYPE 5245				FILM TYPE 5298				5245	5298	TOTAL
	1000' LOADS	400' LOADS	SE		1000' LOADS	400' LOADS	SE		ON HAND	ON HAND	ON HAND
PREVIOUS	4,000	400	0		2,000	2,400	330		4,400	4,730	9,130
TODAY (+)	5,000	2,000	0		3,000	2,000	420		7,000	5,420	12,420
TODAY (-)	1,000	400	0		2,000	0	0		1,400	2,000	3,400
TO DATE	8,000	2,000	0		3,000	4,400	750		10,000	8,150	18,150

Figure 3.44 Completed daily film inventory form #2 for day #2.

PRODUCTION "Claire of the Moon" DATE 10/29/91

DAY # 2

NEG.	ROLL #	AMOUNT	GOOD	NO GOOD	WASTE	TOTAL	SE
5245	4	1000	490	450	60	1000	0
5245	5	400	240	150	10	400	0
5298	6	1000	680	240	80	1000	0
5298	7	1000	370	210	0	580	420
TOTALS		3,400	1,780	1,050	150	2,980	420

	(5245)	(5298)	( )	TOTAL
BALANCE (1)	4,400	4,730		9,130
(+) REC'D TODAY	7,000	5,000		12,000
BALANCE (2)	11,400	9,730		21,130
(-) USED TODAY	1,400	1,580		2,980
BALANCE (3)	10,000	8,150		18,150

Figure 3.45 Completed daily film inventory form #3 for day #2.

Using the information from the above camera reports and inventory forms, the following section breaks down the information and shows where it comes from for each style of daily film inventory form. In examples where information is to be transferred from one day's inventory form to the next day's form, I have included the section from each form for each day.

The following section refers to daily film inventory form #1 in Figure 3.46.

FILM STOCK 5245 DAY #1

ROLL #	GOOD	NO GOOD	WASTE	TOTAL(1)	SE	TOTAL(2)
1	210	150	40	400	0	400
2	280	110	10	400	0	400

Figure 3.46 Breakdown of information for daily film inventory form #1.

- FILM STOCK: The type of film you are using—Kodak 7248, 7293, 5245, 5287; Fuji 8631, 8551; etc. In this example you are using Eastman Kodak Color Negative 5245.
- DAY #: The day number that you are shooting, based on the total number of shooting days and on how many days you have shot previous to today. In this example it is the first day of shooting.
- ROLL #: The camera roll number from the camera report. In this example you have roll numbers 1 and 2.
- GOOD (G): The total of good or printed takes from the camera report for each roll.
- NO GOOD (NG): The total of no good takes from the camera report for each roll.
- WASTE (W): The amount of footage left over that cannot be called a short end. Less than 40 feet in 16mm and less than 100 feet in 35mm is considered waste.
- TOTAL (1): The total of GOOD plus NO GOOD plus WASTE. GOOD + NO GOOD + WASTE = TOTAL (1).

- SE: The amount of footage left over that is too large to be called waste. More than 40 feet in 16mm and more than 100 feet in 35mm is considered a short end.
- TOTAL (2): The total of GOOD plus NO GOOD plus WASTE plus SHORT END. GOOD + NO GOOD + WASTE + SE = TOTAL (2).

The following section refers to daily film inventory form #1 in Figure 3.47.

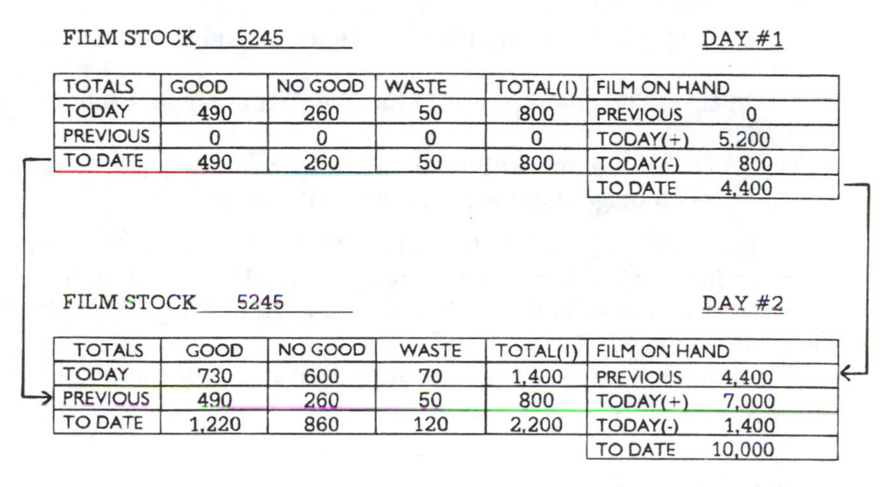

FILM STOCK 5245 — DAY #1

TOTALS	GOOD	NO GOOD	WASTE	TOTAL(1)	FILM ON HAND	
TODAY	490	260	50	800	PREVIOUS	0
PREVIOUS	0	0	0	0	TODAY(+)	5,200
TO DATE	490	260	50	800	TODAY(-)	800
					TO DATE	4,400

FILM STOCK 5245 — DAY #2

TOTALS	GOOD	NO GOOD	WASTE	TOTAL(1)	FILM ON HAND	
TODAY	730	600	70	1,400	PREVIOUS	4,400
PREVIOUS	490	260	50	800	TODAY(+)	7,000
TO DATE	1,220	860	120	2,200	TODAY(-)	1,400
					TO DATE	10,000

Figure 3.47 Breakdown of information for daily film inventory form #1.

- TOTALS: The total amount of all roll numbers combined for each category: GOOD (G), NO GOOD (NG), WASTE (W) and TOTAL (1).
- TODAY: The totals for all roll numbers, shot today for each category: GOOD (G), NO GOOD (NG), WASTE (W), and TOTAL (1). In this example, for day #1, the total good for roll numbers 1 and 2 combined is 490, total no good is 260, total waste is 50, and total (1) is 800.
- PREVIOUS: The totals for all roll numbers shot previous to today, obtained from the previous day's report, from the section labeled Totals—To Date. In this example, for day #1, there are no previous amounts because it is the first day of filming.

- TO DATE: The combined total for all roll numbers shot today plus the totals for all roll numbers shot previous to today. These amounts are then written on the next day's daily inventory report, in the section labeled Totals—Previous.

Film on Hand

- PREVIOUS: The total amount of footage on hand at the start of today for each film stock, obtained from the previous day's report, from the section labeled Film On Hand—To Date. In this example, for day #1 you had no film on hand at the start of the day because it is the first day of filming.
- TODAY (+): The total amount of footage received today for each film stock.
- TODAY (–): The total amount of footage shot today for each film stock.
- TO DATE: The combined total of previous, plus footage received today, less footage shot today, for each film stock.

PREVIOUS + TODAY(+) – TODAY(–) = TO DATE. This is the total amount of footage on hand at the end of the shooting day. This amount is then written on the daily inventory report for the next day in the section labeled Film on Hand—Previous.

The following section refers to daily film inventory form #1 in Figure 3.48.

DAY #1

TOTAL FILM USE - ALL FILM STOCKS

TOTALS	GOOD	NO GOOD	WASTE	TOTAL(!)	TOTAL FILM ON HAND	
TODAY	920	500	50	1,470	PREVIOUS	0
PREVIOUS	0	0	0	0	TODAY(+)	10,600
TO DATE	920	500	50	1,470	TODAY(-)	1,470
					TO DATE	9,130

DAY #2

TOTAL FILM USE - ALL FILM STOCKS

TOTALS	GOOD	NO GOOD	WASTE	TOTAL(!)	TOTAL FILM ON HAND	
TODAY	1,780	1,050	150	2,980	PREVIOUS	9,130
PREVIOUS	920	500	50	1,470	TODAY(+)	12,000
TO DATE	2,700	1550	200	4,450	TODAY(-)	2,980
					TO DATE	18,150

Figure 3.48 Breakdown of information for daily film inventory form #1.

Total Film Use—All Film Stocks

- TOTALS: The total amount of all roll numbers, for all film stocks combined, for each category: GOOD (G), NO GOOD (NG), WASTE (W), and TOTAL (1).
- TODAY: The combined total for today only, for all film stocks, for each category: GOOD (G), NO GOOD (NG), WASTE (W), and TOTAL (1).
- PREVIOUS: The combined total for all film types shot previous to today, for each category: GOOD (G), NO GOOD (NG), WASTE (W) and TOTAL (1). This amount is obtained from the previous day's daily report form from the section labeled Total Film Use—To Date.
- TO DATE: The combined total of all film stocks shot today plus the total of all film stocks shot previous to today. These amounts are then written on the next day's daily inventory report, in the section labeled, Total Film Use—Previous.

Total Film on Hand

- PREVIOUS: The combined total amount of footage on hand at the start of today for all film stocks, obtained from the previous day's report, from the section labeled, Total Film On Hand—To Date.
- TODAY (+): The combined total amount of footage received today for all film stocks.
- TODAY (–): The combined total amount of footage shot today for all film stocks.
- TO DATE: The combined total of previous, plus footage received today, less footage shot today for all film stocks.

PREVIOUS + TODAY(+) – TODAY(–) = TO DATE. This is the total amount of footage on hand at the end of the shooting day. This amount is then written on the daily inventory report for the next day in the section labeled, Total Film on Hand—Previous. Remember, these figures are combined totals for all film stocks on hand during the production.

The following section refers to daily film inventory form #2 in Figure 3.49.

DAY #1

FILM TYPE	LOAD TYPE	ROLL #	GOOD	NO GOOD	WASTE	TOTAL	SE
5245	400	1	210	150	40	400	0
5245	400	2	280	110	10	400	0
5298	1000	3	430	240	0	670	330

Figure 3.49 Breakdown of information for daily film inventory form #2.

- FILM TYPE: The film stock you are using—Kodak 7248, 7293, 5245, 5287; Fuji 8631, 8551; etc. In this example you are using Eastman Kodak Color Negative 5245 and 5298.
- LOAD TYPE: The size of the roll loaded in the magazine. In this example, roll number 1 is a 400 foot roll, roll number 2 is a 400 foot roll, and roll number 3 is a 1000 foot roll.
- ROLL #: The camera roll number from the camera report. In this example, you have roll numbers 1, 2, and 3.
- GOOD (G): The total of good or printed takes from the camera report for each roll.
- NO GOOD (NG): The total of no good takes from the camera report for each roll.
- WASTE (W): The amount of footage left over that cannot be called a short end. Less than 40 feet in 16mm and less than 100 feet in 35mm is considered waste.
- TOTAL (T): The total of GOOD plus NO GOOD plus WASTE. GOOD + NO GOOD + WASTE = TOTAL.
- SE: The amount of footage remaining that is too large to be called waste. More than 40 feet in 16mm and more than 100 feet in 35mm is considered a short end. In this example there was a 330 foot short end created from roll number 3.

The following section refers to daily film inventory form #2 in Figure 3.50.

DAY #1

	GOOD	NO GOOD	WASTE	TOTAL	SE
PREVIOUS	0	0	0	0	0
TODAY (+)	920	500	50	1,470	330
TO DATE	920	500	50	1,470	330

DAY #2

	GOOD	NO GOOD	WASTE	TOTAL	SE
PREVIOUS	920	500	50	1,470	330
TODAY (+)	1,780	1,050	150	2,980	420
TO DATE	2,700	1550	200	4,450	750

Figure 3.50 Breakdown of information for daily film inventory form #2.

- PREVIOUS: The total for all roll numbers in each category shot previous to today: GOOD (G), NO GOOD (NG), WASTE (W), TOTAL (T), and SHORT END (SE). This information is obtained from the previous day's inventory report form, from the section labeled, To Date. In this example, for day #1, there are no previous amounts because it is the first day of filming.
- TODAY (+): The total for all roll numbers shot today in each category: GOOD (G), NO GOOD (NG), WASTE (W), TOTAL (T) and SHORT END (SE). In this example, for day #1 the total good for all roll numbers shot today is 920, total no good is 500, total waste is 50, etc.
- TO DATE: The combined total for each category for all rolls shot previous to today plus all rolls shot today.

PREVIOUS + TODAY(+) = TO DATE. These amounts are then written on the next day's daily inventory report, in the section labeled previous.

The following section refers to daily film inventory form #2 in Figure 3.51.

DAY #1

UN-EXPOSED FILM ON HAND											
	FILM TYPE 5245				FILM TYPE 5298				5245	5298	TOTAL
	1000' LOADS	400' LOADS	SE		1000' LOADS	400' LOADS	SE		ON HAND	ON HAND	ON HAND
PREVIOUS	0	0	0		0	0	0		0	0	0
TODAY (+)	4,000	1,200	0		3,000	2,400	330		5,200	5,730	10,930
TODAY (-)	0	800	0		1,000	0	0		800	1,000	1,800
TO DATE	4,000	400	0		2,000	2,400	330		4,400	4,730	9,130

DAY #2

UN-EXPOSED FILM ON HAND											
	FILM TYPE 5245				FILM TYPE 5298				5245	5298	TOTAL
	1000' LOADS	400' LOADS	SE		1000' LOADS	400' LOADS	SE		ON HAND	ON HAND	ON HAND
PREVIOUS	4,000	400	0		2,000	2,400	330		4,400	4,730	9,130
TODAY (+)	5,000	2,000	0		3,000	2,000	420		7,000	5,420	12,420
TODAY (-)	1,000	400	0		2,000	0	0		1,400	2,000	3,400
TO DATE	8,000	2,000	0		3,000	4,400	750		10,000	8,150	18,150

Figure 3.51 Breakdown of information for daily film inventory form #2.

- FILM TYPE: This section is left blank so that you may fill in the type of film you are using—Kodak 7248, 7293, 5245, 5287; Fuji 8631, 8551; etc.
- 1000′ LOADS: This column is to keep totals for 1000-foot rolls received, and on hand, for each film stock you are using. In this example, on day #1, for film stock 5245, you had no film previous, you received 4000 feet in 1000 foot rolls, you used no 1000 foot rolls, and you have on hand at the end of day #1, 4000 feet of 5245 in 1000 foot rolls.
- 400′ LOADS: This column is to keep totals for 400-foot rolls received, and on hand, for each film stock you are using. In this example, on day #1, for film stock 5245, you had no film previous, you received 1200 feet in 400 foot rolls, you used 800 feet in 400 foot rolls, and you have on hand at the end of day #1, 400 feet of film stock 5245 in 400 foot rolls.
- SE: This column is to keep totals for all short ends created, received, and on hand for each film stock you are using. In this example, on day #1, for film stock 5298, you created one short end, which is 330 feet in length. The remaining blank column is

for you to fill in with any other size rolls you may be using during production.

- ON HAND: These columns are to keep combined totals on hand for each film stock and each roll size you are using. Fill in the blank space at the top of the column with the film stock that you are using. 1000′ LOADS + 400′ LOADS + SE = ON HAND.
- TOTAL ON HAND: The combined total of film stock on hand for all film stocks and roll sizes being used.
- PREVIOUS: The total amount of footage on hand at the start of each day for each film stock and each roll size, obtained from the previous day's report, from the section labeled Unexposed Film On Hand—To Date. In this example, for day #1, for film stock 5245, you had no film previous to day #1.
- TODAY (+): The total amount of footage received today for each film stock and roll size. In this example, for day #1, for film stock 5245, you received 4000 feet in 1000 foot rolls, and you received 1200 feet in 400 foot rolls.
- TODAY (–): The total amount of footage shot today for each film stock and roll size.
- TO DATE: The combined total of previous, plus footage received today, less footage shot today for each film stock, and roll size.

PREVIOUS + TODAY(+) – TODAY(–) = TO DATE. This is the total amount of footage on hand at the end of the shooting day. This amount is then written on the daily inventory report for the next day in the section labeled Unexposed Film on Hand—Previous. Remember to separate the amounts for each film stock and roll size when transferring numbers to the next day's film inventory form.

The following section refers to daily film inventory form #3 in Figure 3.52.

DAY #1

NEG.	ROLL #	AMOUNT	GOOD	NO GOOD	WASTE	TOTAL	SE
5245	1	400	210	150	40	400	0
5245	2	400	280	110	10	400	0
5298	3	1000	430	240	0	670	330
TOTALS		1,800	920	500	50	1,470	330

Figure 3.52 Breakdown of information for daily film inventory form #3.

- NEG: The type of film you are using—Kodak 7248, 7293, 5245, 5287; Fuji 8631, 8551; etc. In this example you are using Eastman Kodak Color Negative 5245 and 5298.
- ROLL #: The camera roll number from the camera report. In this example you have roll numbers 1, 2, and 3.
- AMOUNT: The total amount of footage loaded in the magazine for that roll number. In this example, roll number 1 is a 400 foot roll, roll number 2 is a 400 foot roll, and roll number 3 is a 1000 foot roll. GOOD + NO GOOD + WASTE + SE = AMOUNT.
- GOOD (G): The total of good or printed takes from the camera report for each roll.
- NO GOOD (NG): The total of no good takes from the camera report for each roll.
- WASTE (W): The amount of footage left over that cannot be called a short end. Less than 40 feet in 16mm and less than 100 feet in 35mm is considered waste.
- TOTAL (T): The total of GOOD plus NO GOOD plus WASTE. GOOD + NO GOOD + WASTE = TOTAL.
- SE: The amount of footage left over that is too large to be called waste. More than 40 feet in 16mm and more than 100 feet in 35mm is considered a short end
- TOTALS: The combined total in each category for all roll numbers: AMOUNT, GOOD, NO GOOD, WASTE, TOTAL, and SE.

The following section refers to daily film inventory form #3 in Figure 3.53.

DAY #1

	(5245)	(5298)	( )	TOTAL
BALANCE (I)	0	0		0
(+) REC'D TODAY	5,200	5,400		10,600
BALANCE (2)	5,200	5,400		10,600
(-) USED TODAY	800	670		1,470
BALANCE (3)	4,400	4,730		9,130

DAY #2

	(5245)	(5298)	( )	TOTAL
BALANCE (I)	4,400	4,730		9,130
(+) REC'D TODAY	7,000	5,000		12,000
BALANCE (2)	11,400	9,730		21,130
(-) USED TODAY	1,400	1,580		2,980
BALANCE (3)	10,000	8,150		18,150

Figure 3.53 Breakdown of information for daily film inventory form #3.

- ( ) These columns are left blank for you to fill in with the film stock you are using. In this example you are using Eastman Kodak Color Negative 5245 and 5298.
- TOTAL: The combined totals of previous film on hand, film received today, film shot today, and film on hand at the end of today for all film stocks.
- BALANCE (1): The total amount of film on hand at the start of today for each film stock. In this example for day #1 you had no film on hand at the start of the day because it is the first day of filming. This information is obtained from the previous day's inventory report form, from the section labeled Balance (3).
- (+) REC'D TODAY: The total amount of footage received today for each film stock. In this example, for day #1, you received 5200 feet of film stock 5245 and 5400 feet of film stock 5298.
- BALANCE (2): The combined total of previous film on hand plus amount of footage received today. BALANCE (1) + REC'D TODAY = BALANCE (2).
- (-) USED TODAY: The total amount of footage shot today for each film stock.

- BALANCE (3): The total of footage on hand at the end of today, which is the combined total of the previous amount of footage on hand plus the amount of footage received today less the amount of footage shot today. BALANCE (2) – USED TODAY = BALANCE (3). This amount is then written on the daily inventory report for the next day in the section labeled Balance (1).

Ordering Additional Film Stock

Once you have completed filling out the daily film inventory forms at the end of each shooting day, be sure you have enough film on hand to continue filming. As the film inventory gets low, notify the production office that you need additional film stock. A good rule to follow is to have at least enough film on hand for two or three days of filming. Of course, if it is the last day of filming, you probably will not need to order any additional film. Be especially aware of holidays and weekends during the shooting schedule, because you will not be able to order film on these days. Whenever you receive any additional film stock, remember to record the amounts on the daily film inventory form. If possible, obtain a copy of the packing list that came with the film so that you have proof of how much was sent.

Many times the 2nd A.C. will make identification labels for the magazines each time a new supply of film is received. This saves time later when you are rushing to load magazines. Write the basic information on the labels and place them on each film can. Each time a magazine is loaded, remove the label from the can and place it on the magazine, and fill in the magazine number.

Storage and Care of Motion Picture Film

All motion picture films are manufactured to very high-quality standards, and the proper storage and handling of these films is very important. Motion picture films are sensitive to heat, moisture, and radiation. For short-term storage of less than 6 months, original cans of unopened raw stock should be kept at a temperature of 55°F or lower, and at a humidity level below 60%. For long-term storage of more than 6 months, film should be kept at a temperature of between 0°F and –10°F and at a humidity level below 60%. In addition, it should be kept away from any chemicals or fumes that could cause contamination of the emulsion layers. It should not be stored near any exhaust or heating pipes, or in direct sunlight.

When removing any film stock from cold storage, it must be allowed to warm up before opening the can. Failure to allow the film to reach the proper temperature before opening the can will result in condensation forming on the film, causing spots in your photographic image. Never open a film can immediately after removing it from cold storage. Table 3.3 lists the recommended warm-up times for motion picture films as recommended by Eastman Kodak.

All exposed cans of film should be sent to the laboratory as soon as possible. If there is any reason that exposed film cannot be sent to the lab within a reasonable amount of time, then it should be stored according to the recommendations for unexposed film.

Distribution of Reports

Once all the paperwork is completed, distribute copies to the appropriate departments. The production office should receive a copy of the daily film inventory form. The camera department also should keep a copy of the daily film inventory form.

Most camera reports usually consist of four copies. The top copy is always attached to the film can that is sent to the lab with the exposed film. One copy goes to the editor, one copy to the production office, and one copy is kept by the camera department. You should staple the camera reports to the daily film inventory form for each day, so that it will be easier to answer any questions later. In most cases the production office copy is given to the Second Assistant Director so that he or she may fill out the daily production report.

Filing of Paperwork

As you have discovered from previous sections of this chapter, the camera department requires a lot of paperwork, most of which is filled in by the 2nd A.C. This includes camera reports for each roll, inven-

Table 3.3 Recommended Warm-up Time for Sealed Packages of Motion Picture Film.

FILM FORMAT	WARM-UP TIME
16mm	1 - 1 1/2 Hours
35mm	3 - 5 Hours

tory forms, and so on. You should set up some type of filing system to keep all of the paperwork organized during the production. One good method is to use a small plastic file box or cardboard box with various sections for each type of form or paperwork. Some sections in addition to the daily film inventory forms in your filing system are equipment received, equipment returned, film received, and so on. You should have copies of all packing lists for anything received by the camera department as well as anything returned by the camera department. Many assistants keep some type of record book that is used to keep track of equipment received and returned. Each time you receive or return a piece of equipment, enter the date and description of the equipment in the record book. This way, if there are any questions later, check the book.

Customize your filing system depending on the needs of the particular production. Some 2nd A.C.s are also responsible for keeping time sheets for each member of the camera department. Each day mark down the hours worked by each member of the department, and at the end of the week fill out the time cards. The important thing to remember is to be as complete and as organized as possible so that the production will go smoothly and problems will be minimized.

Performing the Duties of First Camera Assistant

From time to time the 2nd A.C. may be called on to act as 1st A.C. on some shots. There may be an additional camera or the 1st A.C. may have to leave for some reason. It is a good idea to have a basic knowledge and understanding of the job requirements of a 1st A.C. in case this happens. Chapter 4 discusses in detail all the responsibilities of that position.

Packing Equipment

At the end of each shooting day, all the camera equipment should be packed away in its proper case and placed in a safe place until the next shooting day. This should be done as quickly as possible. Remember, the sooner you pack everything away, the sooner you go home for the day. Check all areas of the location to be sure that you have all the camera equipment and nothing is left behind. Place all equipment in the camera truck, or if you are shooting on a stage, place it in a safe area on stage. Many stages have a separate room for the camera department for the storage of equipment. This room also may contain a darkroom for loading and unloading the film. Any camera equipment should be placed in its case and not left out where it could become damaged. This equipment is very valuable and should be handled very carefully. It will be much easier to locate anything if it is put away each time instead of left lying around.

Tools and Accessories

As with many professions, you must have some basic tools and accessories so that you may do the job properly. When first starting out, you should have a very basic tool kit or ditty bag, and as you gain more experience and work more frequently, you can add things as you feel they are needed. Some of the tools are common tools that you may need, while others are specialized pieces of equipment that are unique to the film industry. In addition to the basic tools, an assistant should also have a small inventory of expendables, film cans, cores, camera reports, etc. There will be many times when you are called for a job at the last minute and you may have no time to acquire some of these items. By having a small amount on hand, you will always be prepared for most job calls that you get. See Appendix D for a list of the common tools and equipment that should be included in an assistant cameraman's ditty bag or tool kit.

2nd A.C. Tips

The 1st A.C. must stay close by the camera and the D.P. so that he may assist the D.P. in any way necessary. The 2nd A.C. is there to assist the 1st A.C. by getting equipment when needed, moving equipment for each new setup, and anything else that may be required by the 1st A.C., Camera Operator, or D.P. The camera must never be left unattended, and if the 1st A.C. must step away, the 2nd A.C. will stand by until he returns. Unless the entire cast and crew is on a break, the camera should never be left unattended.

The 1st A.C. and 2nd A.C. are a team and must work together. If you must leave the set for any reason, you should inform the 1st A.C. If the 1st A.C. needs you for something and doesn't know where you are, he may have to leave the camera unattended to take care of the particular matter. The 1st A.C. should also inform you whenever he leaves the set.

Whenever any piece of equipment is called for, you should repeat it back to confirm that you heard the request and that you heard it correctly. If your name is called out, you should also respond so that whoever called will know that you heard him or her.

When changing magazines be sure to enter the new roll number on the ID tape before handing the magazine to the 1st A.C. Never give the 1st A.C. a magazine that does not have an identification label on it and be sure that this label is completely filled in.

When preparing to shoot any scenes, be sure to obtain the proper scene and take numbers from the Script Supervisor. Place this information on the slate so that it is ready when the camera rolls. As soon as the camera cuts, change the take number to the next highest number so you are ready in case the Director decides to shoot the same shot again.

Be prepared to change the scene and take numbers quickly if the shot changes. Keep your eyes and ears open at all times so that you are constantly aware of what is happening on the set. As you become familiar with a particular working style of the 1st A.C. and D.P., you should be able to anticipate their requests and be ready when they do make a certain request. The D.P. may always use a particular lens for the close-up and another lens for the wide shot. By paying attention, you will know when a new scene is being shot, and will have the lens ready when it is called for. Also, pay close attention to what filters are used for certain shots. Watch the D.P. and Director when they are blocking out the scene. This will give you an idea where the camera is to be placed, and it will also be an indication of where you can move the equipment to so that it is close by.

Unless you are told or asked by the 1st A.C., never go into his tool kit, front box, or ditty bag without permission. If something is needed from these, he will either get it personally or tell you that it is OK. You wouldn't like someone using your tools without permission, so treat the 1st A.C. the same way you would like to be treated.

Keep all equipment organized and in its proper place. If it is kept in the same place all of the time, it can easily be located when in a hurry. This applies to both the camera truck and magliner cart. When on stage or location, you should have some type of four-wheel cart (magliner) for moving the equipment from place to place. You will have many equipment cases to deal with each day, and it is much easier and quicker if they can be wheeled from place to place instead of individually carried. Whenever the camera is moved to a new location, the magliner cart should also be moved.

Most important, if you make a mistake, tell the 1st A.C. *immediately*. This information should be communicated to the 1st A.C. quietly so as not to alarm anybody else. It may not be as bad as you think, and between the 1st A.C. and 2nd A.C. you may be able to take care of it without anybody finding out. If you must tell the D.P. or any other production personnel, do it quickly and quietly.

As a 2nd A.C. you must be able to work very closely with the D.P., the Camera Operator, and the 1st A.C. Always maintain a professional attitude, and if you are ever unsure of something, do not be afraid to ask. Always do your job to the best of your ability, and if a mistake is made, admit it so that it can be corrected. Remember that some day you will be in their position, dealing with the same situations and problems.

POST-PRODUCTION

Wrapping Equipment

At the completion of filming, the camera equipment, camera truck, and anything else relating to the camera department must be wrapped. This means that everything should be cleaned and packed away. All equipment must be cleaned, packed, and sent back to the camera rental house. The wrap can take anywhere from a few hours to an entire day, depending on the size of the camera package. Usually the 1st A.C. wraps out the camera equipment, while the 2nd A.C. wraps out the truck, darkroom, and film stock. Many times, if it is a small production, only the 1st A.C. wraps the camera equipment. The truck and darkroom should be left clean for the next job. A final inventory of the film stock should be done, and all film should be packed in boxes for the production company.

4

First Camera Assistant

After two or three years, you probably will move up from Second Camera Assistant (2nd A.C.) to First Camera Assistant (1st A.C.). In Britain and Europe, the 1st A.C. may be called the *Focus Puller*. During production the 1st A.C. works directly with the 2nd A.C., the Camera Operator, and especially the Director of Photography (D.P.). The position of 1st A.C. requires great attention to detail. The 1st A.C. should stay as close as possible to the D.P. during shooting and be prepared for any number of requests. Keeping your eyes and ears open at all times, and never being too far from the D.P. or the camera is a sign of a good 1st A.C. One of the primary responsibilities during shooting is to maintain sharp focus throughout each shot. He is also responsible for the smooth running of the camera department and maintenance of all camera equipment, as well as many other duties. This chapter discusses in detail each of the 1st A.C.'s duties and responsibilities. These duties are separated into three categories: pre-production, production, and post-production.

PRE-PRODUCTION

Choosing Camera Equipment

During pre-production, the D.P. prepares a list of camera equipment that will be needed on the production. Many times she prepares this list with the 1st A.C. Because the 1st A.C. works with the equipment daily, he usually knows which accessories are needed to make the shooting go as smoothly as possible. The D.P. chooses the camera, lenses, and filters, and the 1st A.C. usually determines which accessories are needed to complete the camera package. You should have a working knowledge of all camera systems, the accessories for each, and have copies of rental catalogs from various rental houses to help in choosing the proper equipment. Camera rental houses will give you a copy of their current rental catalog at no charge.

Choosing Expendables

During pre-production make a list of the expendables needed for the camera department. As discussed in Chapter 3, this list is prepared by both the 1st A.C. and the 2nd A.C. Each may have specialty items needed to do their job, which should be included along with the standard items. The standard expendables are items that will be needed in the daily performance of your job, such as camera tape, permanent felt tip markers, ballpoint pens, compressed air, lens tissue, lens cleaning solution, and so on. They are referred to as *expendables* because they are items that are used up or expended during the course of the production. The size and type of production determines which items and how much of each are needed. After the initial order, the 2nd A.C. is responsible for checking the supply on a regular basis to make sure that you do not run out of anything. For a complete list of the standard items on a camera department expendable list, see the Expendables Checklist in Appendix C.

The Rental House

Before going to the camera rental house to prep, you should contact them to be sure that everything is ready for you. They will tell you what time they have scheduled for the prep for your particular production. Once you arrive, all or most of your equipment will be set aside in an area for you to work in. At the rental house you should first check their list against the list that you made with the D.P. to be sure that you have everything. If any item is missing, request it immediately. A technician or prep tech from the rental house will be assigned to you, and this is the person you will communicate with about any problems or questions regarding your equipment.

As an Assistant Cameraman, please be aware that the rental house prep tech's job is not simply to just pull the items requested off a shelf. Before you arrived at the rental house, he has done the same prep, if not a more in-depth prep than the one you will be doing. Each piece of equipment has been thoroughly checked to be sure that it works.

The camera and equipment you have requested will most often be prepared for you and be ready at the time scheduled for the prep. However, remember you are not the only production that the rental house is dealing with at that time, and therefore all of your equipment may not be ready. The rental house prep tech may be working with more than one production company, so you should have a little patience when asking for anything if you do not get it right away. It is always best to schedule your prep a few days ahead and be prepared to work around the rental house's schedule. The rental house will do its best to accomodate your schedule, but sometimes you may have to prep a little earlier or later than planned.

If you need to add any items to your list, be sure that you have the OK from your production company and the rental house, because a deal may have been previously negotiated. Additional equipment may not be part of the original agreement, and the production company needs to know what add-ons you have requested, so that they can authorize it and make additional arrangements with the rental company. If it is an item that is absolutely necessary, the rental house will most often work out an arrangement that is agreeable to all concerned.

As an assistant, it is very important to have a good working relationship with the camera rental house. If you have treated them and their equipment properly in the past, they will be more inclined to help you out when a production company does not have a lot of money, but you need a few additional items. A few years ago I was working for a small production company on a feature film. They had worked out a special price deal for the equipment with the rental house. I asked for a few items that were not included in the original camera package. Because of the excellent relationship that I had with the rental house, the few additional items that I requested were included in the camera package at no additional charge.

A negative attitude will not be tolerated at the rental house. A rental house tech told me of one show that he had prepped where the production company only had a specific amount of money that they could spend for the equipment. The A.C. attempted to get more equipment than the production company was prepared to pay for and was told that he could not have the items. The A.C. then began to throw a fit, claiming these were items that were needed to do the show properly and they must be included. Needless to say, he did not get the items and was not very welcome in the rental house after that.

The rental house prep tech is there to service the needs of the A.C. and production company, but the prep techs are not your personal servants and do not jump when called. The prep tech is there to help, and if you are unfamiliar with a piece of equipment, please ask about it. They would rather spend the time answering the question than fixing it after it was broken through error. But remember, you must be patient as to how quickly your questions are answered because there are other rentals going out at the same time.

Although I have worked in the industry for quite some time and have worked with most of the currently used camera systems, I sometimes forget things. As I have heard it said many times, "If you don't use it, you lose it." I was doing a prep and came across a piece of equipment that I had not used for quite some time. I asked the prep tech to answer some questions about the equipment. He was busy at the time but said he would come back when he could. I continued with the prep and when the prep tech finished what he was doing, he returned to me to answer the questions. If you have forgotten how something works, don't be afraid to ask. In addition, there will very often be times when you are prepping a piece of equipment that you have never used before. You should always ask the rental house prep tech for help on any unfamiliar piece of equipment.

When doing the prep, you should have a system that you follow. Sticking to your system helps to facilitate time and make the prep go quickly and smoothly. However, for various reasons a rental house may not have an item ready, may need to make an exchange, or may even have to get the item from another rental house. You must be flexible and willing to adjust your prep routine if necessary. Remember, you don't have to rush through a prep if you have a scheduled prep time. The rental house will not close on you if they are the ones who cause a delay with a piece of equipment.

PREPARATION OF CAMERA EQUIPMENT

Before you shoot one frame of film, the camera and all related accessories must be checked to be sure that everything is in working order. Once you know when the production will begin filming, and the equipment has been ordered, contact the rental house and arrange a time when you can go and do the camera prep. A camera prep can take anywhere from a couple of hours to one, two, or three days, depending on the size of the production. Be sure to allow enough time to complete the prep so that you are not rushed and are able to check each piece of equipment thoroughly. When you go to the rental house, take along your tools and accessories. Have some dummy loads of film to use for scratch testing the magazines. A *dummy load* is a small spool of film left over from previous shoots and is called waste on the camera report. Instead of throwing it out, many assistants save these short lengths of film for use during the camera prep.

The primary purpose for doing the camera prep is to be sure that you have all the necessary camera equipment and accessories, and that they are in working order. Each item, no matter how small, must be checked and tested. Starting with the spreader, tripod, and head, you will assemble the entire camera package. Each accessory is attached to the camera and tested. Lenses are checked for sharpness and accurate focus, magazines are tested for noise and to be sure that they don't scratch the film, etc. If you find any piece of equipment that is not performing satisfactorily during the prep, send it back immediately and request a replacement.

The following lists the many items found in a typical camera rental package and describe what you should check for during the camera prep. Remember, all items on the following list may not be needed on every production. I have listed the most frequently used items that will be found on most larger productions. All camera packages may not have every item listed below.

Camera Prep Checklist

______ 1. Spreader
Runners slide smoothly and lock in all positions.
Tripod points fit into receptacles.

______ 2. Tripods
Legs slide smoothly and lock in all positions.
Wooden legs are free from cracks and splinters.
Top casting accommodates the head base (flat or bowl).
Always obtain standard tripod and baby tripod.

______ 3. High hat
High hat is mounted on smooth flat piece of wood.
High hat top casting accommodates the head base (flat or bowl).
High hat top casting is usually the same as the tripod top casting.

______ 4. Low hat
Low hat is mounted on smooth flat piece of wood.

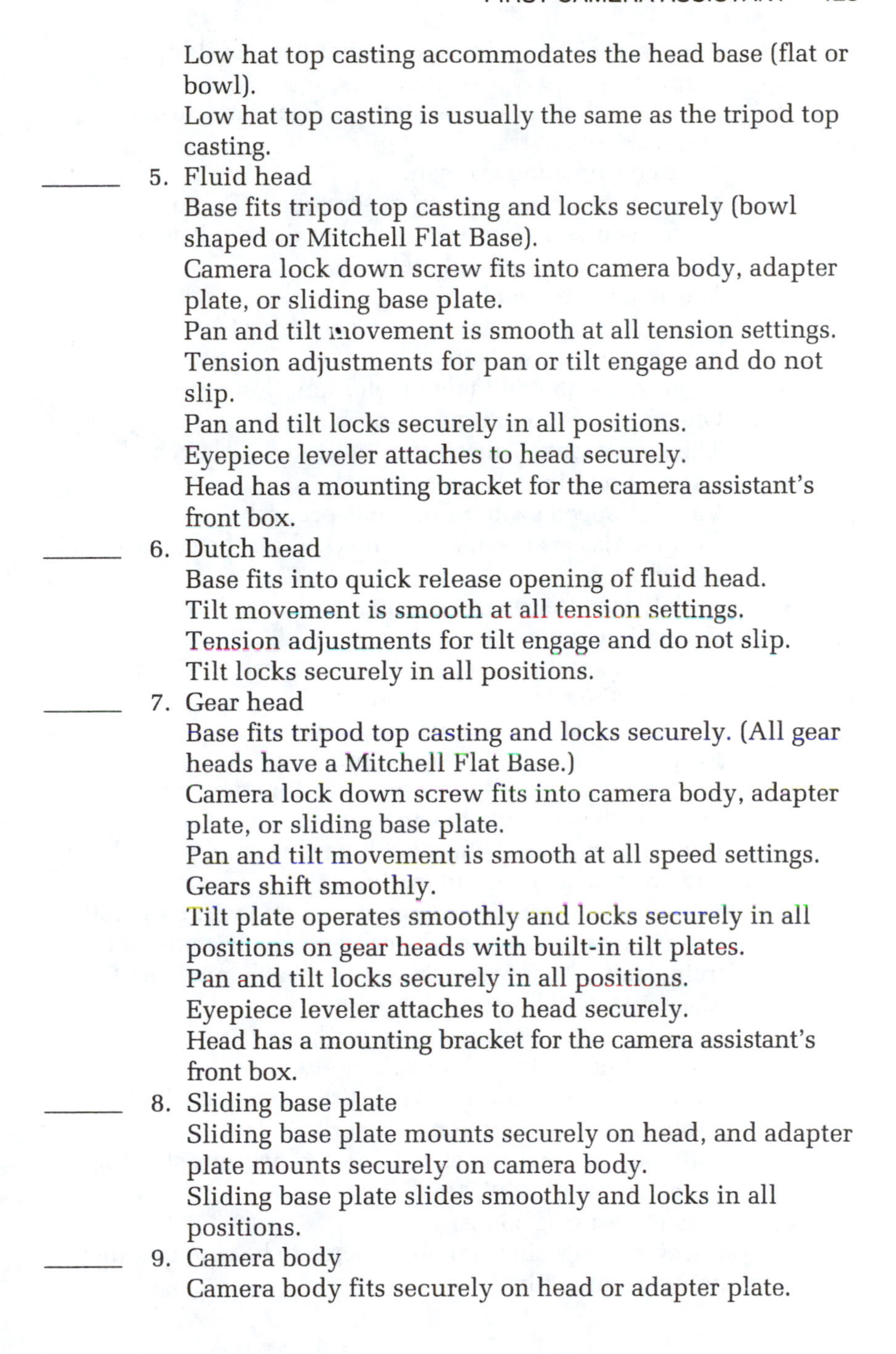

Low hat top casting accommodates the head base (flat or bowl).
Low hat top casting is usually the same as the tripod top casting.

______ 5. Fluid head
Base fits tripod top casting and locks securely (bowl shaped or Mitchell Flat Base).
Camera lock down screw fits into camera body, adapter plate, or sliding base plate.
Pan and tilt movement is smooth at all tension settings.
Tension adjustments for pan or tilt engage and do not slip.
Pan and tilt locks securely in all positions.
Eyepiece leveler attaches to head securely.
Head has a mounting bracket for the camera assistant's front box.

______ 6. Dutch head
Base fits into quick release opening of fluid head.
Tilt movement is smooth at all tension settings.
Tension adjustments for tilt engage and do not slip.
Tilt locks securely in all positions.

______ 7. Gear head
Base fits tripod top casting and locks securely. (All gear heads have a Mitchell Flat Base.)
Camera lock down screw fits into camera body, adapter plate, or sliding base plate.
Pan and tilt movement is smooth at all speed settings.
Gears shift smoothly.
Tilt plate operates smoothly and locks securely in all positions on gear heads with built-in tilt plates.
Pan and tilt locks securely in all positions.
Eyepiece leveler attaches to head securely.
Head has a mounting bracket for the camera assistant's front box.

______ 8. Sliding base plate
Sliding base plate mounts securely on head, and adapter plate mounts securely on camera body.
Sliding base plate slides smoothly and locks in all positions.

______ 9. Camera body
Camera body fits securely on head or adapter plate.

Interior is clean and free from emulsion buildup or film chips.
Aperture plate, pressure plate, and gate are clean and free from any burrs.
Lens port opening is clean.
Mirror is clean and free from scratches. (Do not clean mirror yourself. If it is scratched or dirty, tell someone at the rental house immediately.)
Magazine port opening is clean.
On Panavision cameras, electrical contacts in magazine port openings are clean.
Footage counter and tachometer function properly.
On–off switch functions properly.
The movement of the shutter, pull-down claw, and registration pin is synchronized.
Variable speed switch functions properly.
Ground glass is clean and is marked for the correct aspect ratio.
Variable shutter operates smoothly through its entire range of openings.
Long and short eyepieces mount properly and focus easily to the eye.
Eyepiece heater functions properly.
Eyepiece magnifier functions properly.
Contrast viewing filter on eyepiece functions properly.
Eyepiece leveler attaches to eyepiece securely.
Illuminated ground glass markings function properly and are adjustable in intensity.
Obtain rain covers for all cameras if you will be shooting in any situations where the camera may become wet (rain, snow, in or near any water—beach, pool, etc.)

______ 10. Magazines
Magazines fit securely on camera body.
Doors fit properly and lock securely.
Interior is clean and free from dirt, dust, and film chips.
Footage counter functions properly.
Different size magazines obtained for various shooting situations: 200′, 400′, 1000′, etc.

______ 11. Scratch test magazines
Check all magazines on all cameras to be sure that they do not scratch film.

Load a dummy load of film into each magazine and thread it through the camera.
Run approximately 10–20 feet of film through the camera.
Remove the film from the magazine take-up side and examine it for scratches or oil spots on the base and on the emulsion side. (Turn the film from side to side while looking at it under a bright light. If there are any scratches on either side, they will be noticeable.)
If you find any scratches, request a replacement magazine.
On variable speed cameras, test magazines at various speed settings.

______ 12. Barneys
Obtain the proper size barney for each size magazine.
If necessary obtain a barney for the camera.
On heated barneys check to be sure that the heater functions properly.

______ 13. Lenses
Check that lens seats properly in camera body.
Front and back glass elements are clean and free from scratches.
If any imperfections or scratches are found on the lenses, be sure to notify the rental house personnel immediately.
Iris diaphragm operates smoothly.
Focus gear threads properly.
Focus distance marks are accurate.
On zoom lenses the zoom motor operates smoothly.
Zoom lens tracks properly.
Zoom lens holds focus throughout the zoom range.
Lens shade mounts securely to each lens.
Matte box bellows fits securely around all lenses. If not, obtain various size rubber donuts in order to make a tight seal between lenses and matte box.

______ 14. Zoom lens tracking
Check that the shifting of the image is minimal when zooming in or out.
While looking through zoom lens, line up the crosshairs of the ground glass on the focus chart or on a point in the prep area where you are working.
Lock the pan and tilt so the cross hairs remain centered on the point.

While looking through the camera, zoom the lens in very slowly and then out very slowly, and watch to see if the crosshairs remain centered on the point throughout the length of the zoom. They may shift a small amount, and this is usually acceptable.

If the crosshairs do not remain centered on the point or shift more than just a little, have the rental house check the lens.

______ 15. Power zoom control and zoom motor (Figure 4.1)

Operates smoothly, both zooming in and zooming out.

Variable speed adjustment is accurate.

Camera on/off switch functions properly (if available).

______ 16. Focus eyepiece

With the lens removed, point the camera at a bright light source or white surface.

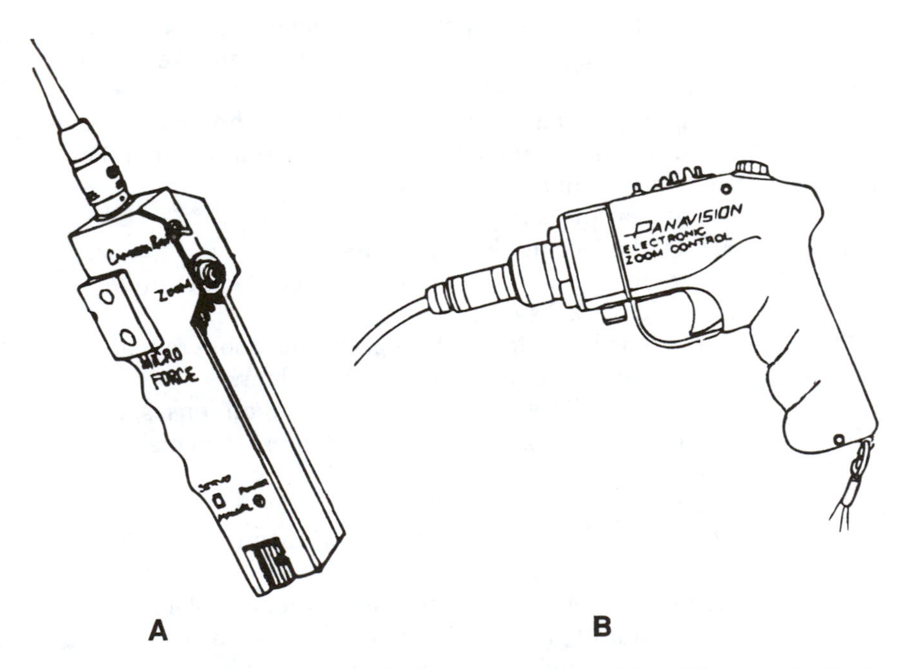

Figure 4.1 (A) Microforce zoom control. (Reprinted from the *Arri SR Book* with permission of the Arriflex Corporation.) (B) Panavision zoom control. (Courtesy of Panavision, Inc.)

While looking through the eyepiece, turn the diopter adjustment ring until the crosshairs are sharp and in focus.
If possible, lock the adjustment ring and mark it so it can be set to the proper position each time you look through the camera.
Wrap a piece of tape around the barrel of the diopter adjustment ring and mark it so that it can be set to the proper mark for each person who looks through the camera.

______ 17. Check focus of each lens
Mount a lens on the camera.
Set the aperture to its widest opening (lowest t-stop number).
Using your tape measure, place a focus chart at various distances from the camera (see Figure 4.2).
At each distance look through the viewfinder eyepiece and focus the chart by eye.
Compare the eye focus to the distance measured and see if they match.

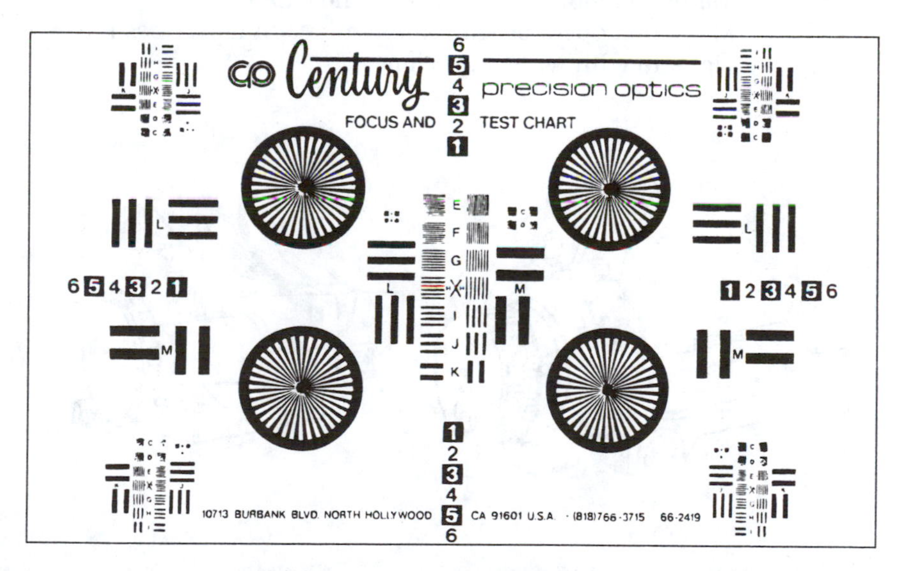

Figure 4.2 Example of focus test chart. (Courtesy of Century Precision Optics.)

If the eye focus does not match the measured focus, have the lens checked by the rental house lens technician.
Check each lens at various distances, including the closest focusing distance and infinity (∞). Check it at one foot intervals up to ten feet and at two foot intervals from ten feet to twenty feet.
With zoom lenses, check the focus with the lens zoomed in all the way (its longest or tightest focal length).
Wide angle lenses do not have to be checked at as many distances as telephoto lenses.

______ 18. Follow focus mechanism (Figure 4.3)
Mount follow focus mechanism securely on the camera or support rods, and be sure it operates smoothly with each lens.
If necessary obtain different focusing gears for prime lenses and zoom lenses.
Check to be sure that you have all accessories, and that they mount and operate properly (whip, speed crank, right hand extension, focus marking disks, etc.).

______ 19. Matte box (Figure 4.4)
Matte box mounts securely on the camera.
Matte box operates smoothly with each lens; does not vignette with wide angle lenses.

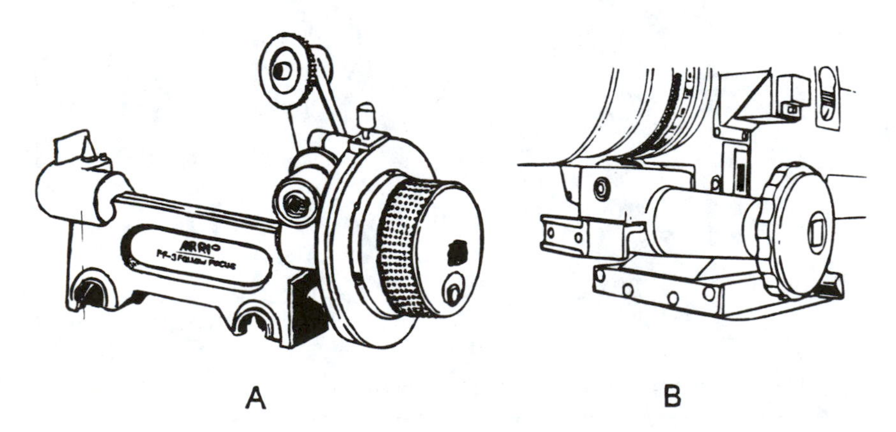

Figure 4.3 (A) Arriflex follow focus mechanism. (Courtesy of the Arriflex Corp.) (B) Panavision follow focus mechanism. (Courtesy of Panavision, Inc.)

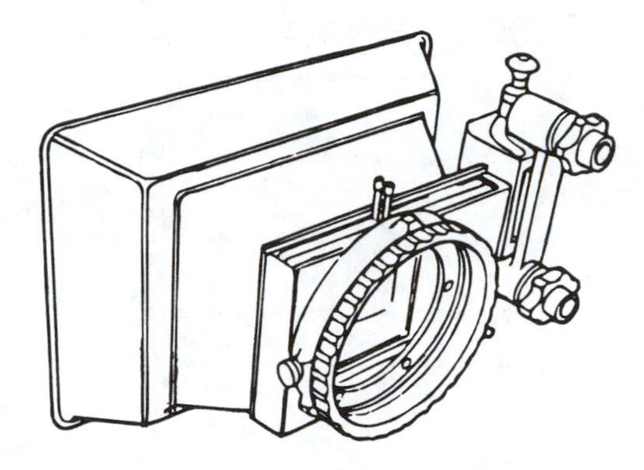

Figure 4.4
Matte box. (Courtesy of Panavision, Inc.)

Matte box has proper adapter rings and rubber donut or bellows for each lens.
Filter trays are the correct size for filters being used.
Filter trays slide in and out smoothly and lock securely in position.
Rotating filter trays or rings operate smoothly and lock securely in position.
Eyebrow mounts securely and can be adjusted easily.
Hard mattes mount securely and are the correct size for each lens.

______ 20. Filters
Each filter is clean and free from scratches.
Filters are proper size for filter trays or retainer rings.
Rotating polarizer operates smoothly.
Always obtain an optical flat or clear filter with any filter set.
Obtain complete sets of filters for all cameras when using more than one camera.
Are graduated filters hard edge or soft edge?

______ 21. Obie light
Obie light mounts securely and operates correctly at each setting.

______ 22. Lens light (assistant's light)
Lens light mounts securely and operates properly.
Lens light is supplied with spare bulbs.

______ 23. Precision speed control (Figure 4.5)
Check that it operates correctly for both high speed and slow motion by running film through the camera at various speeds.

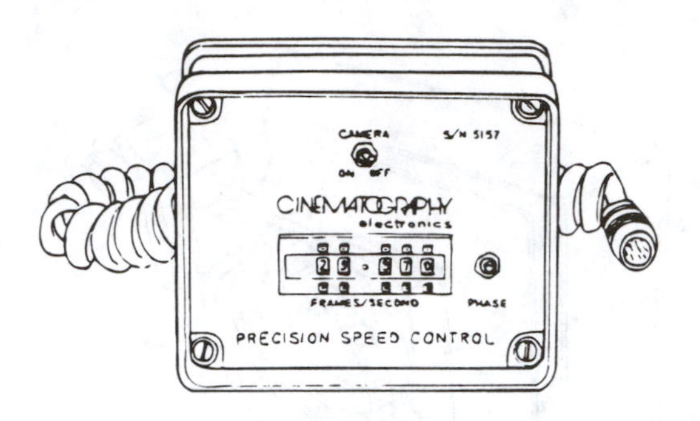

Figure 4.5
Cinematography electronics precision speed control. (Reprinted from the *ARRI 35 Book* with permission of the Arriflex Corporation.)

______ 24. HMI speed control

When using HMI lights be sure that the camera has an HMI speed control so that you may adjust the speed to the correct number when filming. (HMI lights can cause the image to flicker if the camera is not run at certain speeds or shutter angles.)

See Figures 4.6–4.9 for HMI filming speeds and shutter angles.

CINEMATOGRAPHY inc.
electronics
HMI FLICKERFREE FILMING SPEEDS
AT ANY SHUTTER ANGLE
60 Hz Line Frequency

		5.714
		5.454
	12.000	5.217
	10.909	5.000
120.000	10.000	4.800
60.000	9.231	4.000
40.000	8.571	3.750
30.000	8.000	3.000
24.000	7.500	2.500
20.000	7.058	2.000
17.143	6.666	1.875
15.000	6.315	1.500
13.333	6.000	1.000

ALWAYS USE FILM TESTS TO VERIFY RESULTS.

Figure 4.6
HMI filming speeds at any shutter angle, 60 Hz line frequency—United States. (Courtesy of Cinematography Electronics, Inc., Santa Monica, CA.)

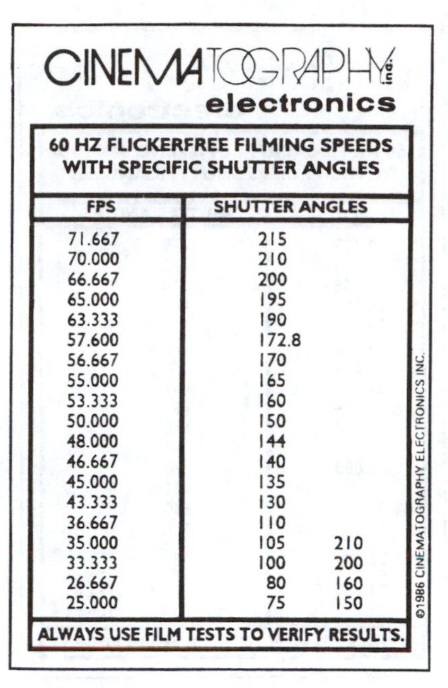
CINEMATOGRAPHY inc.
electronics

60 HZ FLICKERFREE FILMING SPEEDS
WITH SPECIFIC SHUTTER ANGLES

FPS	SHUTTER ANGLES	
71.667	215	
70.000	210	
66.667	200	
65.000	195	
63.333	190	
57.600	172.8	
56.667	170	
55.000	165	
53.333	160	
50.000	150	
48.000	144	
46.667	140	
45.000	135	
43.333	130	
36.667	110	
35.000	105	210
33.333	100	200
26.667	80	160
25.000	75	150

ALWAYS USE FILM TESTS TO VERIFY RESULTS.

©1986 CINEMATOGRAPHY ELECTRONICS INC.

Figure 4.7
HMI filming speeds at specific shutter angles, 60 Hz line frequency—United States. (Courtesy of Cinematography Electronics, Inc., Santa Monica, CA.)

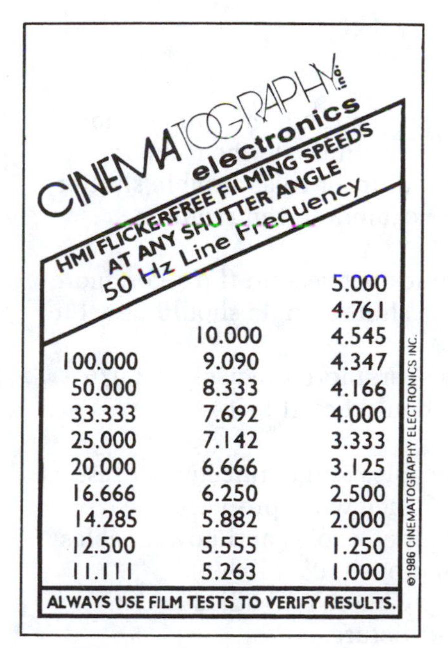
CINEMATOGRAPHY inc.
electronics

HMI FLICKERFREE FILMING SPEEDS
AT ANY SHUTTER ANGLE
50 Hz Line Frequency

		5.000
		4.761
	10.000	4.545
100.000	9.090	4.347
50.000	8.333	4.166
33.333	7.692	4.000
25.000	7.142	3.333
20.000	6.666	3.125
16.666	6.250	2.500
14.285	5.882	2.000
12.500	5.555	1.250
11.111	5.263	1.000

ALWAYS USE FILM TESTS TO VERIFY RESULTS.

©1986 CINEMATOGRAPHY ELECTRONICS INC.

Figure 4.8
HMI filming speeds at any shutter angle, 50 Hz line frequency—Europe. (Courtesy of Cinematography Electronics, Inc.,

CINEMATOGRAPHY inc.
electronics

50 HZ FLICKERFREE FILMING SPEEDS WITH SPECIFIC SHUTTER ANGLES		
FPS	SHUTTER ANGLES	
59.722	215	
58.333	210	
55.556	200	
54.167	195	
52.778	190	
48.000	172.8	
47.222	170	
45.833	165	
44.444	160	
41.667	150	
40.000	144	
38.889	140	
37.500	135	
36.111	130	
30.556	110	
29.167	105	
27.778	100	
24.000	86.4	172.8
22.222	80	160
ALWAYS USE FILM TESTS TO VERIFY RESULTS.		

Figure 4.9
HMI Filming speeds at specific shutter angles, 50 Hz line frequency—Europe. (Courtesy of Cinematography Electronics, Inc., Santa Monica, CA.)

______ 25. Sync box
When shooting TV screens, computer monitors, and projectors, use a sync box to eliminate the roll bar.
If possible, the camera should have a variable shutter so that you can sync the camera to the monitor or screen.
When shooting at 30 frames per second (f.p.s.) or, more precisely, 29.97 f.p.s., the shutter angle should be set to 180°.
When shooting at 24 f.p.s. or, more precisely, 23.976 f.p.s., the shutter angle should be set to 144°.

______ 26. Video tap and monitor
Check that you have all cables and connectors necessary for the video tap and that they work properly.
Have various lengths of video cables and power cables for various shooting situations (10′, 25′, 50′).
Connect video monitor to camera and adjust video camera to obtain the best picture.

______ 27. Handheld accessories

If the production involves any handheld shots, be sure you have the necessary accessories, which should attach securely and operate properly (left- and right-hand grips, shoulder pad, handheld follow focus, clamp-on matte box or lens shade, 400′ or 500′ magazines, on-board batteries, etc.).

Connect the hand grip with an on–off switch and be sure that it operates properly.

______ 28. Remote start switch

Connect remote start switch to camera and be sure that it operates properly.

Be sure to have long enough cable if necessary for different shooting situations (dangerous shots, car shots, stunts, etc.).

______ 29. Batteries and cables

All cables should be in good condition and have no frayed or loose wires.

At least two battery cables should be obtained for each camera being used.

There should be no loose pins in the plugs.

Battery cables should be of various lengths for different shooting situations.

At least two batteries should be obtained for each camera being used.

Extra batteries should be available for each camera in case you will be shooting high speed.

If you will be shooting any handheld shots, you should have at least two battery belts or on-board batteries.

Each battery should have a charger, either built-in or separate.

See Figures 4.10 and 4.11.

______ 30. Camera tests

At the end of the camera prep, the D.P. may ask you to shoot some tests. You usually need only one 400′ roll of film for these tests, depending on how involved the tests are. Be sure that you have a light meter available to obtain the correct exposure.

A. Film registration test

Check that the registration of the camera is accurate by filming a registration test chart (Figure 4.12).

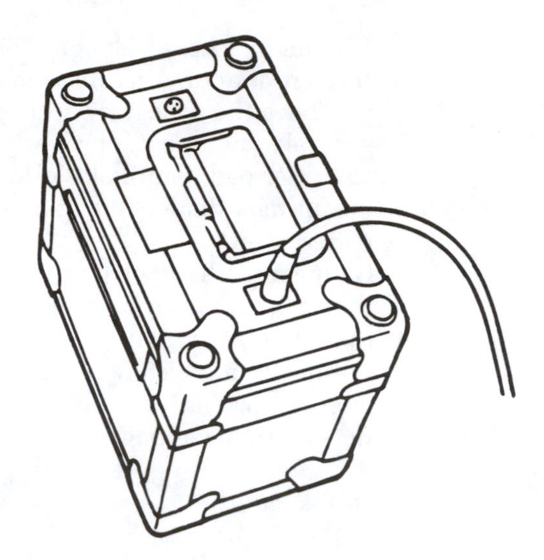

Figure 4.10
Panavision 24 volt double block battery. (Courtesy of Panavision, Inc.)

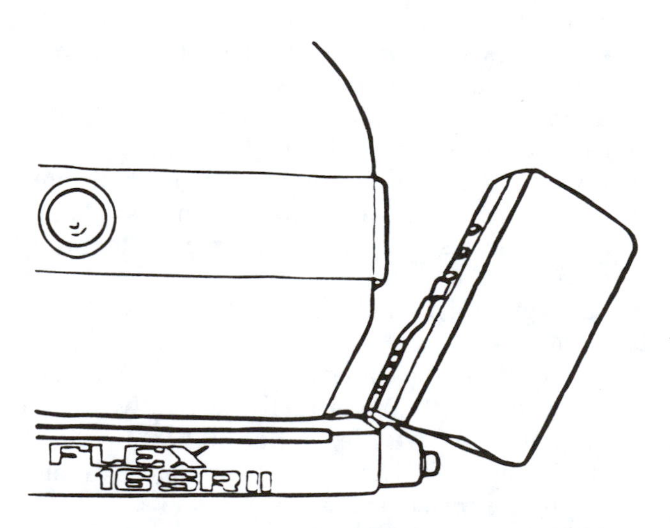

Figure 4.11 Arriflex 16SR camera with on-board battery. (Reprinted from the *16SR Book* with permission of the Arriflex Corporation.)

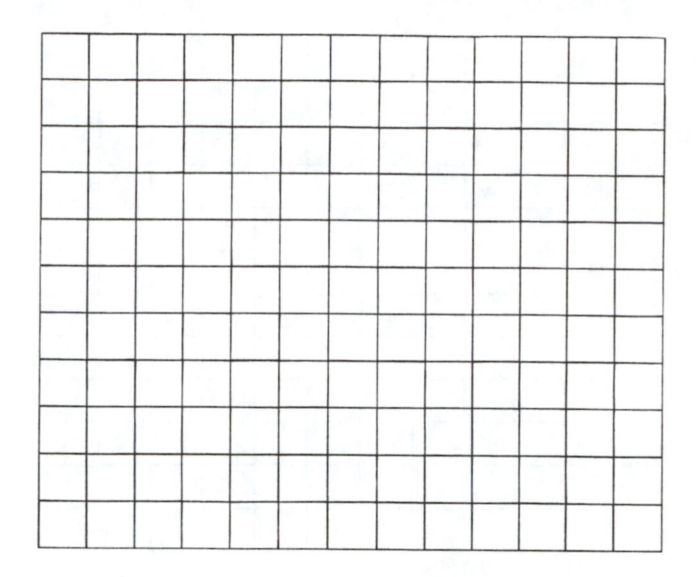

Figure 4.12 Example of registration test chart.

Thread film in camera and mark the exact frame where you start, using a permanent felt tip marker (Figure 4.13).
Line up the registration test chart through the eyepiece so that the crosshairs of the ground glass are centered on the lines of the chart (Figure 4.14).
Lock the pan and tilt on the head.
Shoot approximately 30 feet of the chart at normal exposure.

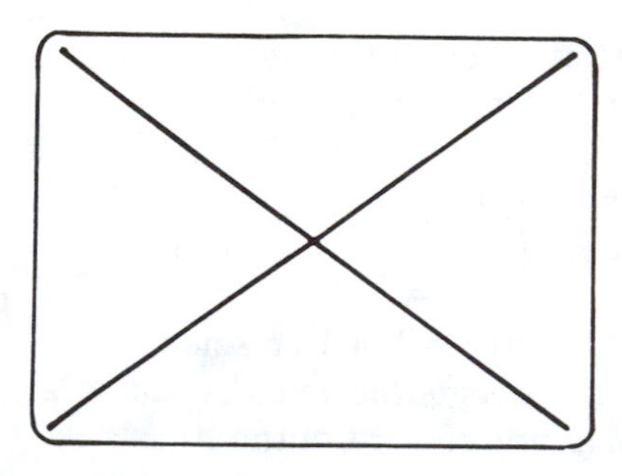

Figure 4.13
How to mark the starting frame before shooting the registration test.

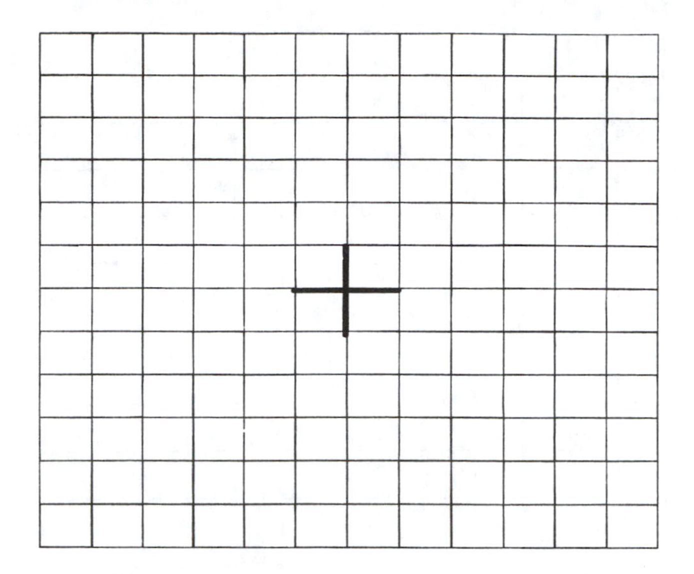

Figure 4.14 Positioning of crosshair on registration chart for shooting the first exposure.

Remove the magazine, do not break the film, and go into the darkroom or use a changing bag.
Wind the film back to the starting frame that you marked. You may have to wind the film back to the very beginning of the roll in order to find the start frame. (I recommend shooting the registration test first on the roll, because it is easier to wind the film back to the beginning of a roll than to a place in the middle.)
Place the magazine back on the camera, and thread the film so that your original mark is again lined up in the gate.
Release the pan and tilt of the head, and reposition the camera to line up the registration chart through the eyepiece so that the crosshairs of the ground glass are centered within one of the boxes of the chart (Figure 4.15).
Lock the pan and tilt on the head.
Shoot approximately 50 feet of the chart at one stop underexposed.
When the film is projected, there will be two sets of chart lines on the screen. If the registration of the camera is correct, there should be no movement of the lines.

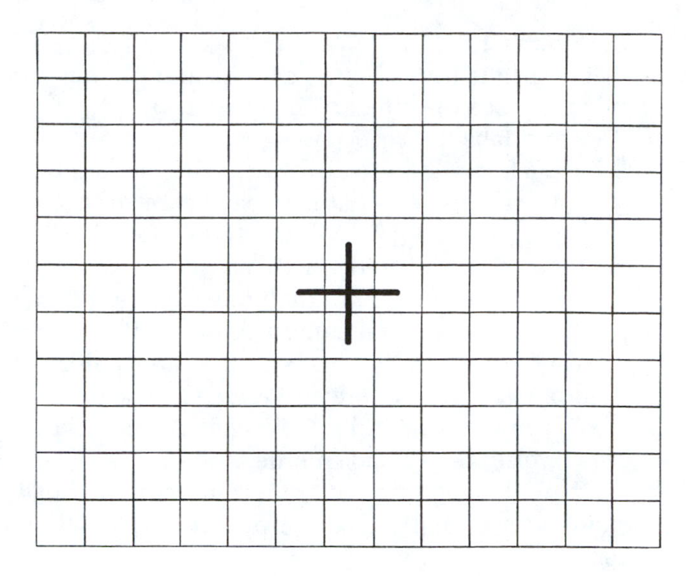

Figure 4.15 Positioning of crosshair on registration chart for shooting the second exposure.

B. Lens focus calibration test
Place each lens on the camera and frame the focus chart so that the entire chart is framed by the lens.
Photograph the focus test chart at various distances to be sure that the lens maintains sharp focus.
Make sure the image is sharp in the center, as well as on the left and right sides of the frame and the top and bottom of the frame.
Shoot the focus test with the lens open to its widest aperture whenever possible.
C. Lens color balance test
Place each lens on the camera and photograph a color chart to check that there is consistent color balance between lenses.
Check to see that each lens reproduces all colors the same way.
If you do not have a color chart, cut a variety of color photographs out of magazines and paste them on a sheet of poster board and shoot this for the color balance test.

D. Filter test
Place various filters on the camera and photograph a live subject to see the effect of each filter.

______ 31. Pack and label all equipment
Label each case on top and sides with camera tape (just a brief description of what is in each case so that you can find things quickly: CAMERA, 1000′ MAGS, FILTERS, PRIMES, ZOOM, HEAD, AKS, etc.).
If you will be using more than one camera on the production, label each camera case and its corresponding accessories with the same color tape. (For example, all "A" camera and accessory cases would be labeled in red tape and all "B" camera and accessory cases would be labeled in blue tape.)
If you will be working out of a camera truck, label the shelves for each piece of equipment the same as the cases.
For a definition of each item on the camera prep check list, see the glossary.

When doing a camera prep, be as thorough as possible, and check every little item in each equipment case. This is important not only for when you are shooting but also when you do the camera wrap at the end of shooting. If you have checked everything completely and it is listed on the original order, there should be no questions when the equipment is returned to the rental house.

A camera prep is not always a guarantee that nothing will go wrong with the equipment. An experience I had a few years ago illustrates this. I was hired to work on a commercial that was to shoot for two days around Los Angeles. The camera prep took me a couple of hours the day before we were scheduled to start shooting. The next day's call time was 5:00 a.m., and the location was about a one-hour drive away. When I arrived at the location, I proceeded to set up the camera. I connected the battery and turned on the camera. Nothing happened. The camera would not run. After checking all of the batteries and power cables, I telephoned the rental house at their 24-hour number. When I explained to the camera technician what was happening, he instructed me to change one of the internal fuses in the camera. After changing the fuse, the camera still would not run, so it was decided that the production company would return to L.A., get another camera, and shoot the scenes in the studio that were scheduled for the next day. After exchanging the camera, everything went smoothly. I later found out that there was an additional fuse in the camera that could only be accessed and changed by a camera technician, and it was this fuse that had blown out, causing the camera not to run. This is an excellent example of why you should do a camera prep. It also shows that a camera prep is not always a guarantee that something won't go wrong. Without doing the camera prep, I would have had no way of knowing if the camera had been in working order when it was picked up.

PRODUCTION

Setting Up the Camera

The first thing you should do each day is set up the camera. Place the camera on the head, which is either mounted to the tripod or the dolly. If the camera and head are being placed on a tripod, many assistants will use a piece of carpet in place of the spreader under the tripod. The points on the tripod legs dig into the carpet, creating a firm support for the tripod and camera. This also sometimes makes it easier when moving or re-positioning the camera. This piece of carpet is usually three feet by three feet or four feet by four feet in size.

If necessary, oil the camera movement and clean the gate and aperture plate to remove any dirt, dust, or emulsion buildup. Warm up the camera before placing any magazine on it. A general rule to follow is to run the camera for approximately the length of the first roll of film that you will be placing on the camera. For example, if the first roll being placed on the camera is 400′, reset the footage counter to zero and then warm up the camera until the footage counter shows "400." While the camera warms up, attach the remaining accessories to it, including the follow focus, matte box, eyebrow, eyepiece leveler, lens light, and so on. You should not place a lens on the camera or remove a lens from the camera while it is warming up. The shutter turns during the warm-up, and it may hit the back of the lens if the lens is not placed on the camera properly. While the 1st A.C. sets up the camera, the 2nd A.C. stands nearby, handing pieces of equipment to him.

Unless the D.P. requests a specific lens, place a wide or normal focal length lens on the camera. This is to allow the D.P. or Camera Operator to see as much of the scene as possible when they first look through the camera. Open up the aperture to its widest opening and set the focus to the approximate distance so that the scene can be viewed clearly. Once the camera is warmed up, place the first magazine on it and finish making it ready for the first shot. Be sure to let the D.P. and Camera Operator know as soon as the camera is ready for use. This set up procedure should take approximately 15 to 20 minutes from start to finish. It is important to get the camera set up as quickly as possible at the start of each day, but never trade safety for speed. In other words, set it up quickly but don't go so quickly that you could make mistakes.

Keep any camera equipment that will be needed throughout the shooting day as close to the camera as possible without being in the way of other people or equipment. Camera equipment can include lenses, filters, magazines, and accessories. Many assistants have some type of hand cart or dolly to keep camera equipment cases on, which enables the assistant to keep all of the cases neat and organized, and also allows him to move them quickly when there is a new camera setup. The most common type of cart or dolly used for moving camera equipment cases is the Magliner Gemini Jr. It collapses for shipping and storage, and can be set up quickly when needed (Figures 4.16 and 4.17).

Most assistants will also keep their tools and accessories near the camera during shooting. The camera assistant's ditty bag, containing basic tools and accessories, should also be kept on the cart during filming.

Loading and Unloading Film in the Camera

Whenever a new magazine is placed on the camera, notify the D.P. and Camera Operator so that they know that the camera will not be available to them while you complete the reload. Reloading the camera with film should only take a minute or two. Before a new magazine is placed on the camera, clean the interior of the camera body with compressed air. Check and clean the gate and aperture plate to remove any emulsion buildup. If possible, remove the gate and aperture plate for cleaning. When cleaning the gate, never use any type of sharp tool that could scratch it and cause scratches on the film emulsion. To clean emulsion buildup use an orangewood stick, which should have been ordered with the expendables. Clean the gate and aperture plate with compressed air.

When the new magazine is placed on the camera, reset the footage counter to zero so that the dial readings and footage amounts

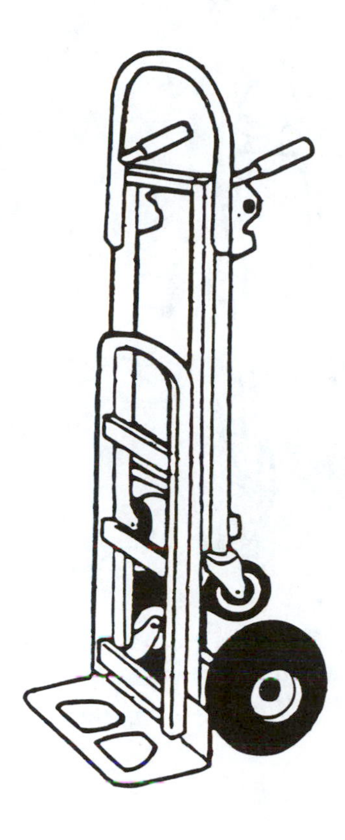

Figure 4.16
Magliner cart folded for storage and transporting.

on the camera reports will be accurate. Remember to write the roll number on the identification label of the magazine if it has not already been done by the 2nd A.C. If the magazine contains a short end, place the small identification label next to the footage counter as a reminder that the magazine does not contain a full roll of film. In addition, when loading a short end, the size and weight of the roll could affect the balance of the camera. Check and rebalance the camera if necessary so that the operator will not have difficulty in operating the shot because of an unbalanced camera. Also, when using a camera with displacement magazines, as the film travels through the camera it is displaced from the front of the camera to the back. This will also cause the camera to become unbalanced, so you should periodically check the balance during shooting and adjust it as required. If necessary, place a barney on the magazine after it has been placed

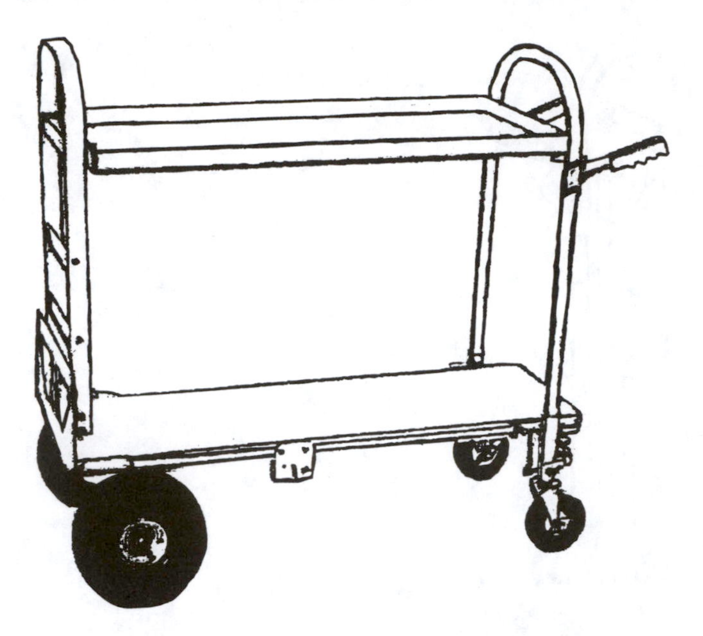

Figure 4.17 Magliner cart with top and bottom shelves, set up for use.

on the camera. Once you have completed loading the magazine and film on the camera, notify the D.P. and Camera Operator that the camera is now ready for use.

Keeping the Camera Clean

One of the most important things to remember during filming is to keep the camera clean. Dirt and dust on the camera not only looks unprofessional, but it also can cause big problems if it gets into the camera body, the magazines, on lenses or filters, in the gate, or on the shutter mirror. The smallest speck of dirt can cause emulsion scratches on the film and ruin a whole day's shooting.

Clean the camera each day when it is set up. Clean the inside with compressed air. Keep the outside of the camera body clean by using an inexpensive two-inch wide brush to brush off the dust and dirt. If the exterior of the camera body becomes exceptionally dirty, wipe it off with a damp cloth. Never use the damp cloth to clean lenses or filters. As stated earlier, clean the gate using an orangewood stick and compressed air. When oiling the camera movement, remove any excess oil by using a cotton swab or foam-tip swab.

Each time a new lens or filter is placed on the camera, check it for dirt, dust, or smudges. If the lens needs to be cleaned, wait until the D.P. or Camera Operator has approved the change and then remove it to clean it before the shot. The proper way to clean lenses and filters will be explained later in the section on lenses.

Oiling the Camera

The movement in many motion picture cameras must be oiled at regular intervals. If you are not sure whether you should oil the movement or how often to oil the camera movement, always check with the rental house when you do the camera prep. The rental house also should give you a small container of oil. In addition to oiling the movement, it is sometimes necessary to lubricate the pull-down pins with a small drop of silicone liquid to prevent squeaking.

Some cameras need to be oiled every day; others only require oiling on a weekly or monthly basis. Check with the rental house, because you can do just as much damage by oiling too much as by not oiling enough. Panavision says that, as a general rule, their cameras should be oiled on a daily basis depending on how much film is being shot each day. When using the Panavision high-speed cameras, they must be oiled after every 1000 feet of film has been shot, whenever you are filming at speeds greater than 60 f.p.s. Panavision cameras have anywhere from seven to thirteen oiling points in the movement, depending on which model you are using. They usually have an oiling diagram on the inside of the door to the camera body. Arriflex cameras use a different type of movement and do not need to be oiled nearly as often. Some Arriflex cameras require oiling only every few months or after a specified amount of film has been run through them.

When you do oil the movement, it is necessary to place only one drop of oil on each oiling point. Be very careful not to get any oil in the gate or on the mirror. If the oil does get onto the film, it will show up as spots on the exposed negative. If you should get any excess oil in the movement, remove it by using a cotton swab or foam-tip swab from the expendables. Be very careful when using the cotton swabs so that you do not leave any of the lint from the cotton tip in the movement. If lint gets into the gate, it can cause hairs on the emulsion. If you do find it necessary to place a drop of silicone liquid on the pull-down claw, or sometimes on the aperture plate, be extra careful not to get any of the silicone in the movement because it can damage it.

Another thing to remember when oiling the camera is to only use the oil supplied by the rental house or recommended for that particular camera. Never use Panavision oil on Arriflex cameras or Arriflex oil on Panavision cameras. It is a good idea to have a supply of the different oils in your kit with your tools and accessories. This way, if you do not get any oil from the rental house, you will be prepared and be able to oil the camera movement when necessary. Figures 4.18–4.23 show the oiling points for some of the cameras that require oiling most often.

Remember: Never over oil the camera movement. Use only the supplied or recommended oil for a particular camera. When in doubt, check with the rental house.

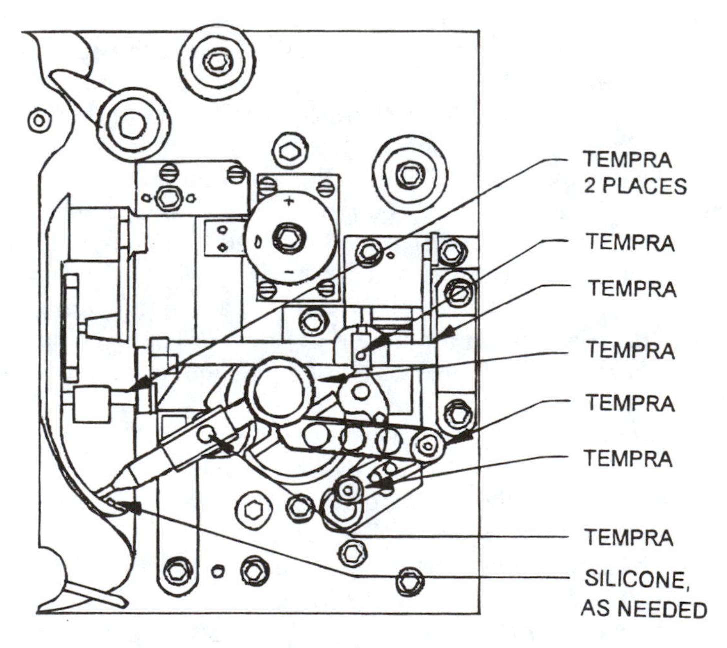

Figure 4.18 Leonetti Ultracam oiling points. (Courtesy of the Leonetti Company.)

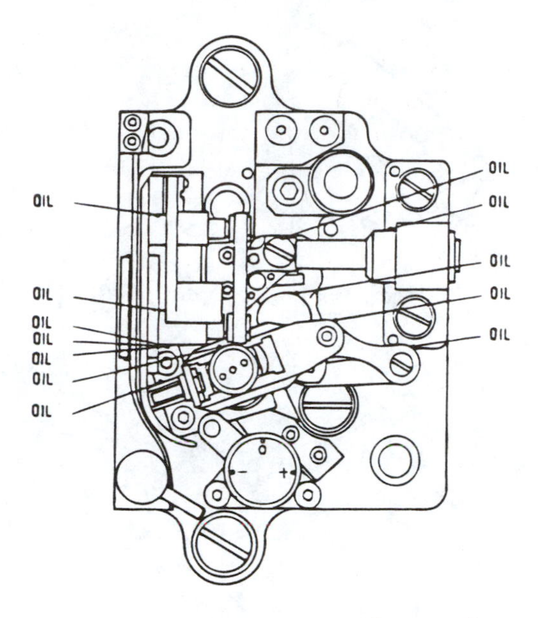

Figure 4.19 Panavision Panaflex 16 oiling points. (Courtesy of Panavision, Inc.)

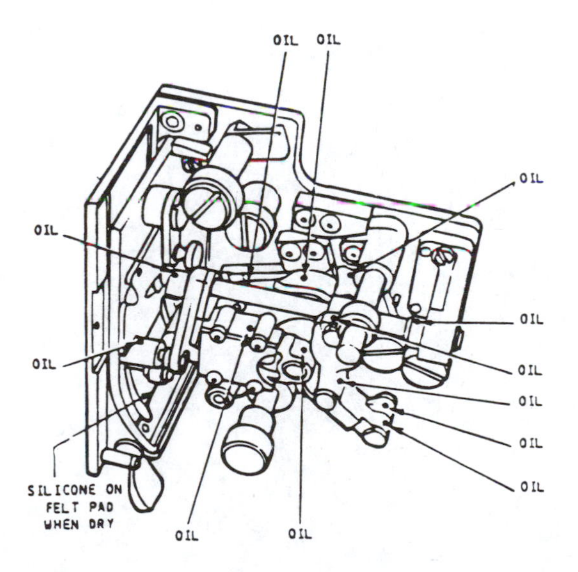

Figure 4.20 Panavision Panaflex 35 oiling points. (Courtesy of Panavision, Inc.)

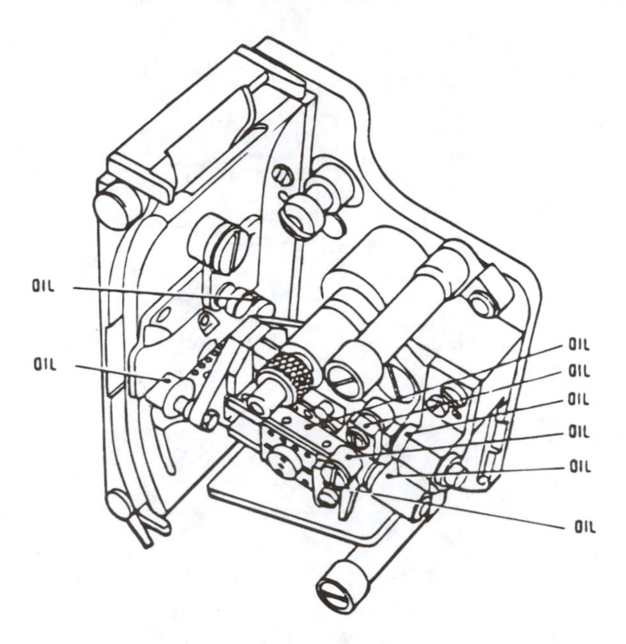

Figure 4.21 Panavision Panastar oiling points. (Courtesy of Panavision, Inc.)

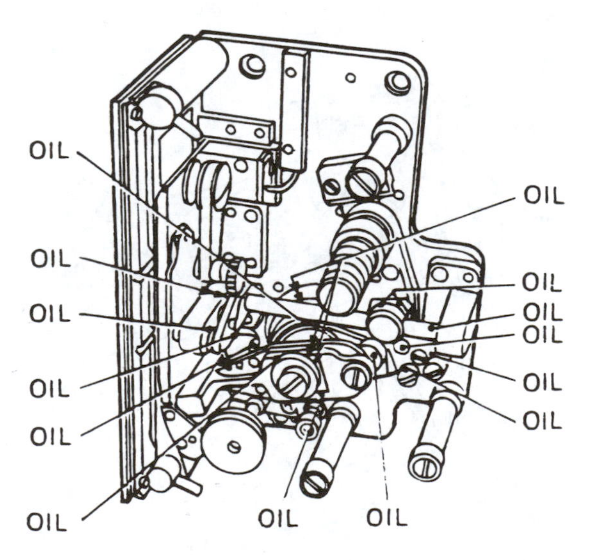

Figure 4.22 Panavision Super PSR oiling points. (Courtesy of Panavision, Inc.)

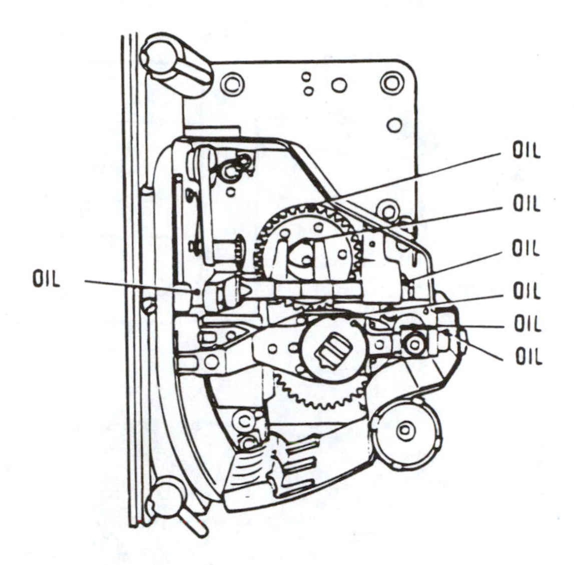

Figure 4.23 Panavision 65mm Camera oiling points. (Courtesy of Panavision, Inc.)

Setting the Viewfinder Eyepiece

The viewfinder eyepiece must be set for each key person who looks through the camera. On most productions the key people are the Director, D.P., Producer, Camera Operator, and 1st A.C. On commercials it may also be set for the client or agency people. Because each person's vision is different, you will need different settings on the eyepiece so that the image will appear sharp and in focus to each person looking through it.

To focus the eyepiece, first remove the lens. Aim the camera at a bright light source or white surface. While looking through the eyepiece, turn the diopter adjustment ring until the crosshairs of the ground glass appear sharp and in focus. Figure 4.24 shows the frame lines and crosshair that is seen when looking through the viewfinder.

Wrap a thin piece of tape around the diopter ring so that it is marked for each person's setting. Figure 4.25 shows the viewfinder marked for the key people who may look through the camera.

Each time one of the key people looks through the camera, set the viewfinder to his or her mark. Remember to set it back to the Camera Operator's mark before shooting.

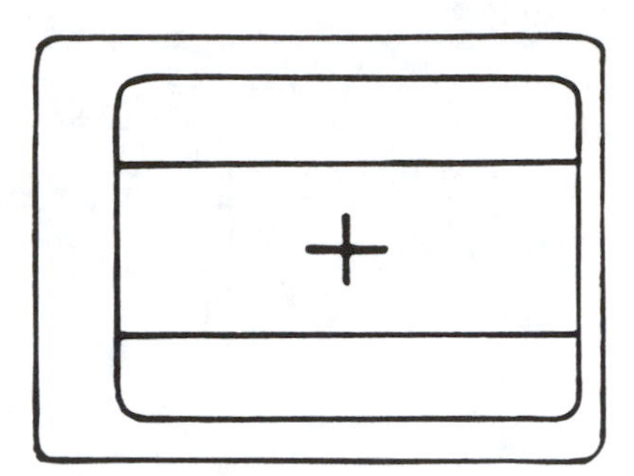

Figure 4.24
View through the eyepiece showing the frame lines and crosshair on the ground glass.

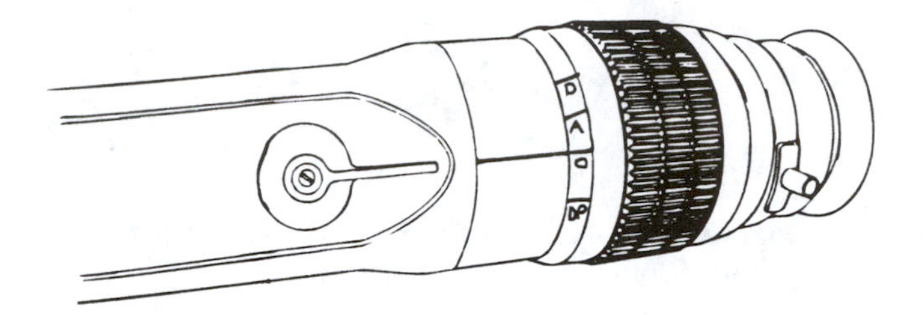

Figure 4.25 Eyepiece showing marks for each person's setting. D = Director, A = 1st A.C., O = Camera Operator, DP = Director of Photography.

To focus the eyepiece while the lens is still on the camera, first be sure that the aperture is set to its widest opening. Look through the lens and adjust the focus until everything is out focus. Aim the camera at a bright light source or white surface. Turn the diopter adjustment ring until the crosshairs of the ground glass appear sharp and in focus. If you wear eyeglasses, it is recommended to remove your glasses before setting the viewfinder eyepiece for your vision.

Checking for Lens Flares

Each time the camera or lights are moved to a new position, you should check that no lights are kicking or shining directly into the lens, which will cause a flare in the photographic image. A flare causes an overall washing out of the image, so objects in the scene don't have sharp detail. If you have a French flag or eyebrow on the camera, you may be able to adjust it to eliminate the flare (Figures 4.26 and 4.27).

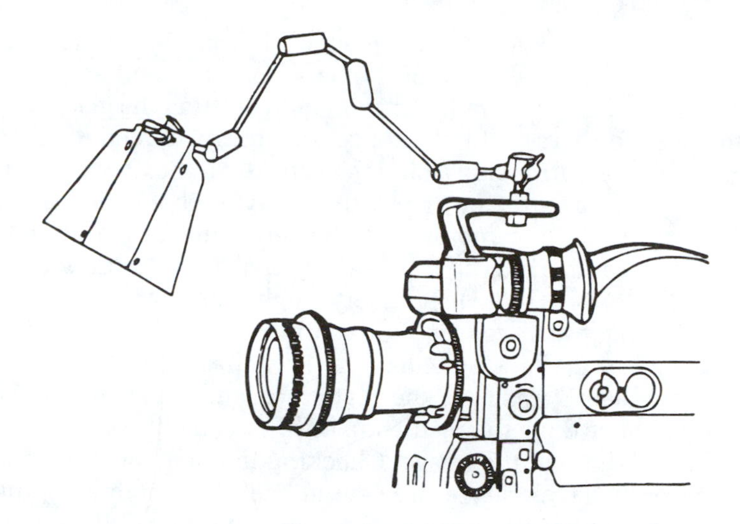

Figure 4.26 French flag in place on camera to eliminate lens flare. (Reprinted from the *16SR Book* with permission of the Arriflex Corporation.)

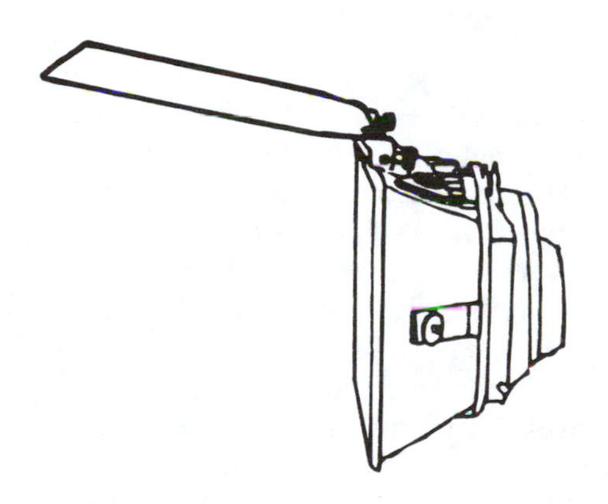

Figure 4.27
Eyebrow in place on matte box to eliminate lens flare. (Courtesy of Panavision, Inc.)

You also may be able to eliminate the flare by placing a hard matte on the matte box. If the flare cannot be removed at the camera, request that a flag be set by one of the grips to keep the light from kicking.

There are a few different ways to check for lens flares. One is to place your face directly next to the lens, looking in the direction that the lens is pointed. Look around the set to see if any lights are shining directly at you, which means they are shining directly into the lens.

Another way to check for flares is to stand in front of the camera and look at the front glass element of the lens from different angles to see if there are any highlights or kicks from lights hitting the glass. A third way to check for flares is to place a convex mirror directly in front of the lens with the mirror side facing the set. Any lights that are causing a flare can be seen in the mirror. A final way to check for flares is to stand in front of the camera, face the lens, and move your hand around the lens or matte box. If you see a shadow from your hand falling across the lens there is probably a light flaring the lens from the angle of the shadow. If you find a flare, it must be removed from the front element of the lens by either the French flag, eyebrow, or grip flag. Remove lens flares from the matte box and filter as well as the lens. Any light striking the matte box or filter can still reflect into the lens causing a flare in the image. Checking the lens for flares takes a certain amount of experience and cannot be fully explained or understood unless you are in an actual shooting situation. Whenever you are not sure if there is a lens flare, ask one of the grips to double-check for you.

Lenses

The basic definition of a lens is that it is an instrument that bends light waves in such a way to produce an image of the object that the light was reflected from. The lens directs the reflected light from an object onto the film emulsion, producing a photographic image of the object. When referring to a lens, the D.P. will call for it by its focal length. The focal length of the lens is an indication of how much of the scene the lens will see. A lens with a short focal length, such as 18mm, 24mm, etc., will see a bigger area of the scene than a lens with a long focal length, such as 85mm, 100mm, 150mm, etc. The focal length of the lens is always expressed in millimeters. There is an exception to this, which is most commonly used by some older D.P.s who have worked in the industry for many years. A 25mm lens may be referred to as a one-inch lens, 50mm as two-inch, 75mm as three-inch, and 100 mm as four-inch. This is based on the fact that twenty-five millimeters is approximately equal to one-inch.

The two main types of lenses are prime and zoom. *Prime lenses* have a single focal length, such as 25mm, 35mm, 50mm, 65mm, and so on. *Zoom lenses* have varying focal lengths, which means that you can change the focal length by turning the barrel of the lens. Zoom lenses are available in many different ranges including 10–100,

20–100, 12–120, 25–250, 150–600, and so on. The 10–100, 25–250, and 12–120 ranges are referred to as ten to one (10--1) zooms. The 20–100 range is called a five to one (5–1) zoom, and the 150–600 is called a four to one (4–1) zoom, and so on. These abbreviated names for the lenses are equal to the ratio of the tightest focal length of the lens to its widest focal length. The zoom lens sizes mentioned are only a small sampling of the different zoom lenses available today. Check with the rental house to see what size zooms they have. Another type of lens that you may be using is a telephoto lens. The *telephoto lens* is also a prime lens but it has a very long focal length, such as 200mm, 300mm, 400mm, 600mm, and even 1000mm. Lenses may generally be classified as wide angle, normal, or telephoto. This is in reference to the area of the scene that they see.

Whenever a lens is not being used it should be capped on both the front and rear elements and placed in a padded case. The padding will help to cushion the lens and protect it from shocks and vibration. The internal elements of the lens can become loosened very easily if the lens is not protected or handled properly. When you are filming in dusty conditions or any situation in which something may strike the front of the lens, it is a good idea to use an optical flat. An *optical flat* is simply a clear piece of glass placed in front of the lens as a means to protect it. Optical flats are available in the same standard sizes as filters. It is much less expensive to replace a filter that has become scratched than to replace the front element of the lens.

All professional lenses have a coating on the front element and should only be cleaned when it is absolutely necessary. Clean a lens first with compressed air or some type of blower bulb syringe. If there are no smudges or fingerprints, then there is nothing more that you need to do to clean the lens. If the lens has any fingerprints or smudges, clean it with lens cleaner and lens tissue. After the dirt and dust have been blown away, moisten a piece of lens tissue with lens cleaning fluid. Wipe the surface of the lens carefully, using a circular motion. While the lens is still damp from the lens solution, use another piece of tissue to remove the remaining lens cleaning fluid from the lens. I have also seen some assistants apply the lens cleaning fluid directly to the lens element. I don't recommend this, because of the curvature of the front element of the lens. The fluid can travel along the element of the lens, and sometimes gets between the lens housing and the glass element ending up behind the element. As a result, you have no way to remove the fluid from the back of the lens glass. The important thing to remember is that you should *never* use a dry piece of lens tissue on a dry lens surface. Small particles of dirt and dust may still be on the coating and will cause scratches. Also, *never* use any type of silicone-coated lens tissue or cloth to clean lenses.

Filters may be cleaned by using the same method for cleaning lenses. First clean the filter with compressed air, and then use lens cleaner and lens tissue. Another good way to clean the filter is by breathing on it and wiping off the moisture with a piece of lens tissue.

Remember: Clean lenses only when absolutely necessary. Never use a dry piece of lens tissue on a dry lens. Never use silicone-coated lens tissue or cloth to clean a lens.

Depth of Field

Depth of field is defined as the range of distance within which all objects will be in acceptable sharp focus, including an area in front of and behind the principal point of focus. There will always be more depth of field behind the principal point of focus than in front of it (Figure 4.28).

Depth of field is determined by the following three factors. You must know all of these in order to determine your depth of field.

1. Focal length of the lens
2. Size of the aperture (f-stop)
3. Subject distance from the camera film plane

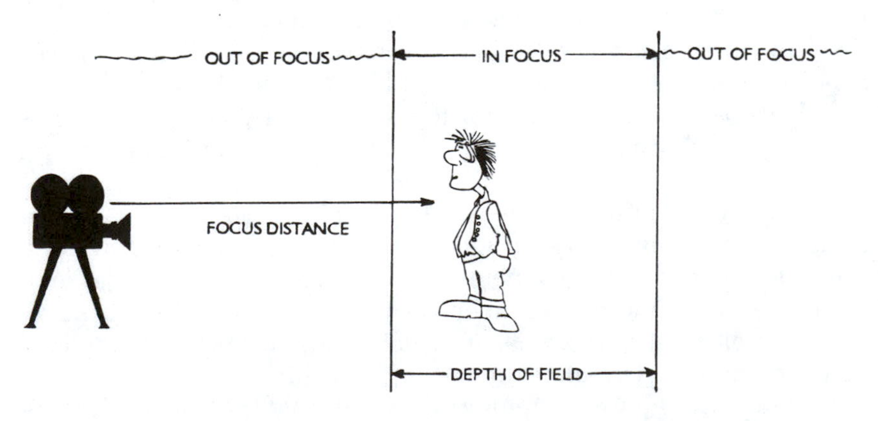

Figure 4.28 Diagram illustrating the principles of depth of field.

In order to find your depth of field for a particular situation, you may use the depth-of-field tables available in many film books. There are depth of field tables available for a variety of focal length lenses. An example of a depth-of-field table is shown in Figure 4.29.

For this table our focal length is 50mm. Let's use an aperture of 2.8, and a distance of 15 feet to determine our depth of field. Knowing these three things enables you to read that the depth of field is from 13 feet, 4 inches to 17 feet, 3 inches. When expressing your depth of field, it is always stated as a range from the closest point to the farthest point and not as a single number. When using depth-of-field tables, remember that the depth of field is different depending on whether you are working with 16mm or 35mm.

LENS FOCAL LENGTH = 50mm **CIRCLE OF CONFUSION = .001"**

	f 1.4	f 2	f 2.8	f 4	f 5.6	f 8	f 11	f 16	F 22
LENS FOCUS DISTANCE	NEAR FAR	NEAR FAR	NEAR FAR	NEAR FAR	NEAR FAR	NEAR FAR	NEAR FAR	NEAR FAR	NEAR FAR
3'	3' 3' 1"	2' 11" 3' 1"	2' 11" 3' 1"	2' 11" 3' 1"	2' 10" 3' 2"	2' 10" 3' 3"	2' 9" 3' 4"	2' 8" 3' 6"	2' 6" 3' 9"
4'	3' 11" 4' 1"	3' 11" 4' 1"	3' 11" 4' 2"	3' 10" 4' 3"	3' 10" 4' 4"	3' 8" 4' 5"	3' 7" 4' 8"	3' 5" 5'	3' 2" 5' 6"
5'	4' 11" 5' 1"	4' 10" 5' 2"	4' 10" 5' 3"	4' 9" 5' 4"	4' 8" 5' 6"	4' 6" 5' 9"	4' 4" 6'	4' 1" 6" 8"	3' 9" 7' 7"
6'	5' 10" 6' 2"	5' 10" 6' 3"	5' 9" 6' 4"	5' 7" 6' 6"	5' 6" 6' 8"	5' 3" 7' 1"	5' 7' 7"	4' 8" 8' 7"	4' 3" 10' 2"
7'	6' 10" 7' 3"	6' 9" 7'4"	6' 7" 7' 5"	6' 6" 7' 8"	6' 3" 8'	6' 8' 6"	5' 9" 9' 2"	5' 3" 10' 9"	4' 10" 13' 6"
8'	7' 9" 8' 4"	7' 8" 8' 5"	7' 6" 8' 7"	7' 4" 8' 11"	7' 1" 9' 4"	6' 9" 10'	6' 4" 11'	5' 9" 13' 4"	5' 2" 17' 8"
9'	8' 8" 9' 4"	8' 7" 9' 6"	8' 4" 9' 9"	8' 2" 10' 2"	7' 10" 10' 8"	7' 5" 11' 7"	6' 11" 13'	6' 4" 16' 4"	5' 8" 23' 8"
10'	9' 7" 10' 5"	9' 5" 10' 8"	9' 3" 10' 11"	8' 11" 11' 5"	8' 7" 12' 1"	8' 1" 13' 4"	7' 6" 15' 3"	6' 9" 20'	5' 11" 32'
12'	11' 5" 12' 8"	11' 2" 13'	10' 11" 13' 5"	10' 6" 14' 1"	10' 15' 2"	9' 4" 17' 1"	8' 7" 20' 5"	7' 8" 30'	6' 9" 67'
15'	14' 1" 16' 1"	13' 9" 16' 6"	13' 4" 17' 3"	12' 8" 18' 5"	12' 20' 4"	11' 23' 11"	10' 30' 10"	8' 9" 59'	7' 7" INF
20'	18' 5" 21' 11"	17' 10" 22' 10"	17' 1" 24' 2"	16' 1" 26' 7"	14' 11" 30' 8"	13' 6" 39' 10"	12' 63'	10' 2" INF	8' 7" INF
25'	22' 7" 28' 1"	21' 8" 29' 7"	20' 7" 31'11"	19' 1" 36'	17' 5" 44'	15' 5" 66'	14' 168'	11' INF	9' INF
50'	41' 1" 64'	38' 72'	35' 88'	31' 131'	27' 376'	22' INF	19' INF	14' INF	11' INF

Figure 4.29 Depth-of-field table: focal length of lens = 50mm, film format = 35mm.

Many assistants also use some type of depth-of-field calculator that allows them to dial in the focal length, f-stop or t-stop, and subject distance and then read the depth of field. Some types of depth-of-field calculators are the Guild Kelly Calculator for both 16mm and 35mm, and the Samcine Mark II Calculator.

The following examples illustrate how the three factors that determine depth of field affect it.

1. Size of the aperture or f-stop: You have more depth of field with larger f-stop numbers (smaller aperture openings) than with small f-stop numbers (larger aperture openings).

> *Example:* A large aperture, such as f 2.8, has less depth of field at a specific distance than at a small aperture, such as f 8, at the same distance (Figure 4.30).

Figure 4.30 Diagram illustrating how the size of aperture (f-stop) affects depth of field.

2. Focal length of the lens: You have more depth of field with wide angle lenses than with telephoto lenses at the same f-stop and distance.

> *Example:* A wide-angle lens, such as a 25mm, will have more depth of field at a specific distance and f-stop than a telephoto lens, such as a 100mm, at the same distance and f-stop (Figure 4.31).

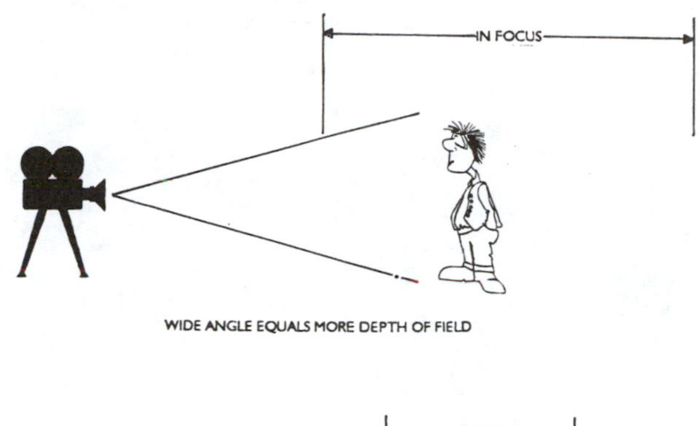

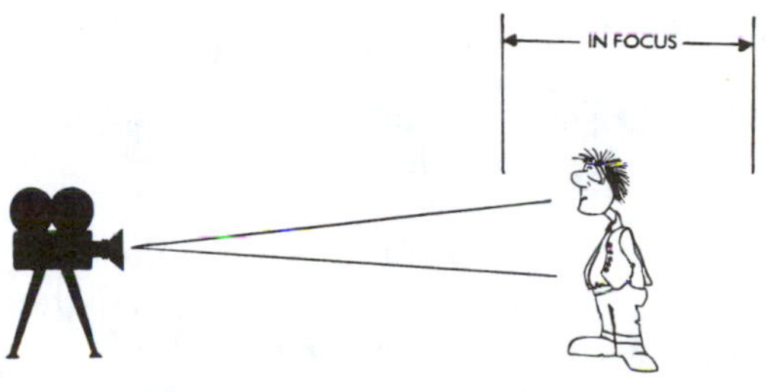

Figure 4.31 Diagram illustrating how the focal length of the lens affects depth of field.

3. Subject distance from the camera: You have more depth of field with a distant subject than with a close subject at the same f-stop and focal length.

> *Example:* An object 20 feet from the camera at a specific f-stop and focal length has more depth of field than an object 8 feet from the camera at the same f-stop and focal length (Figure 4.32).

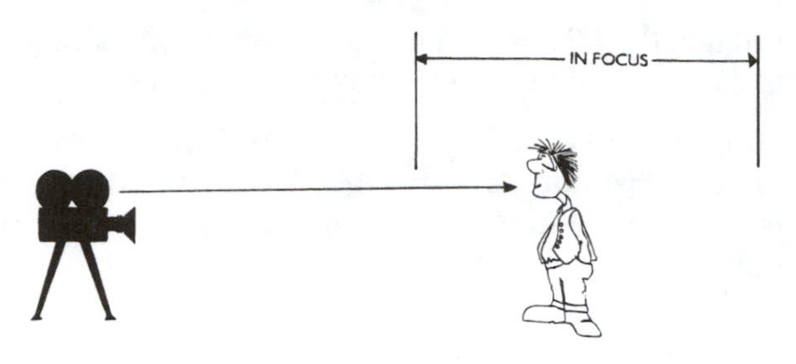

Figure 4.32 Diagram illustrating how the subject distance affects depth of field.

There is a special case of depth of field when you must keep two objects at different distances from the camera in focus in the same shot. When holding focus on two different objects in the same scene, one closer to the camera than the other, you do not set the focus at a point halfway between the two. Because of the principles of depth of field, you should focus on a point that is one third of the distance between the two objects, in order to have both of them in focus. This principle or theory is usually called the *one-third rule*. Remember that this is a theory that does not work in every situation. You should check the depth-of-field tables or use the depth-of-field calculators to be sure. It will depend on the depth of field for the one-third point. Check to see if the distance to each object falls within this range. If it does, then the one-third principle works. If not, you may have to move the objects, change lenses, change the lighting, or only keep one of them in focus at a time. Also, check with the D.P. about whether you should split the focus or whether you should favor one actor over another in the scene. The following example illustrates the one-third rule.

Example: The first object is 6 feet from the camera, and the second is 12 feet from the camera. Using the one-third principle, set the focus at 8 feet in order to have both objects in focus. The distance between the two is 6 feet (12 – 6 = 6). One third of this distance is 2 feet (6 ÷ 3 = 2). Set the focus at 8 feet (6 + 2 = 8). Using the depth of field table from figure 4.29, we see that this example will only work for f-stop numbers of 16 or higher (Figure 4.33).

Figure 4.33 Diagram illustrating the one-third principle for splitting focus between two objects.

Another special case of depth of field is something called hyperfocal distance. The *hyperfocal distance* is defined as the closest point in front of the lens that is in acceptable focus when the lens is focused at infinity (∞). You must check the depth-of-field tables to find out your hyperfocal distance for a given focal length and f-stop. If you set the focus of the lens to the hyperfocal distance, your depth of field will be from one-half the hyperfocal distance to infinity. In other words, setting the focus to the hyperfocal distance gives you the maximum depth of field.

When calculating the depth of field always use f-stops. Depth-of-field tables and calculators base all depth-of-field calculations on f-stop numbers and not t-stop numbers. Use the t-stop only when setting the aperture on the lens.

Don't take the depth-of-field calculations too literally. The focus does not fall off abruptly at the near and far range of depth of field. It is more of a gradual decrease to where a point that is sharp and in focus becomes a blurred circle that is out of focus.

F-Stops and T-Stops

In professional cinematography all lenses may be calibrated in both f-stops and t-stops. The difference between the two is that an f-stop is a mathematical calculation, based on the size of the diaphragm opening, and the t-stop is a measurement of the actual amount of light transmitted through the lens at each diaphragm opening. The f-stop is determined by dividing the focal length of the lens by the diaphragm opening. The f-stop doesn't accurately represent the amount of light coming through the lens because it doesn't take into account the amount of light loss caused as the light passes through the various glass elements within the lens. All exposure readings are given in f-stops. Because the t-stop is an actual measurement, it is more accurate and should always be used when setting the exposure on the lens. In referring to the exposure readings and aperture settings, most camera people will use the terms f-stop and t-stop interchangeably.

When the D.P. gives you the exposure reading for a particular shot, repeat it back to him. This reminds him of what he told you and also enables him to change it if necessary. Most D.P.s try to maintain a constant exposure, especially on interior locations, so if they forget to give you an exposure reading, you probably will be safe if you set the aperture to the setting of the previous shot. Check with the D.P. for each new setup to be sure that you set the correct exposure.

If for some reason you forget to set the exposure, or you set the wrong exposure, notify the D.P. immediately. He will then request another take of the shot with the exposure set correctly. It is much more professional to admit the mistake at the time it was made than to try to cover it up. If you do not let the D.P. know about the error, it will be discovered when the dailies are viewed and the shot comes up on the screen either underexposed or overexposed.

When setting the stop on the lens, you should open the lens to its widest opening and then close down to the correct stop. This will compensate for any sticking that may occur in the leaves of the diaphragm if you just changed from one stop to another.

Example: You are using a lens that has a widest opening of f 1.4. The lens is currently set at f 5.6. The D.P. instructs you to change the stop to a f 4. Open the lens all the way to f 1.4 and then stop down to the new setting of f 4.

As mentioned in Chapter 1, the standard series of f-stop or t-stop numbers is

1, 1.4, 2, 2.8, 4, 5.6, 8, 11, 16, 22, 32, 45, . . .

Each number represents one full f-stop and each full stop admits one-half as much light as the one before it. For example, f 4 admits one-half as much light through the lens as f 2.8. Figure 4.34 shows examples of f-stop numbers and the corresponding diaphragm openings.

Figure 4.34 Diaphragm openings for different f-stop or t-stop settings.

There is certain terminology that is used to refer to the changing of the f-stop opening. When we say *stopping down* or *closing down* the lens, it means that the diaphragm opening will get smaller and the numbers larger. *Opening up* the lens means that the diaphragm opening gets larger and the numbers smaller. *Increasing the stop* is the same as opening up the lens, and *decreasing the stop* is the same as stopping down the lens. When you change from one f-stop number to a larger number (smaller opening), you are closing down or stopping down the lens. When you change from one f-stop number to a smaller number (larger opening), you are opening up the lens. Opening up the lens by one stop will double the amount of light striking the film, and closing down by one stop will cut that amount of light in half.

Example: The current aperture setting is T 5.6. Stopping down or decreasing the stop makes the aperture become T 8. Opening up or increasing the stop makes the aperture become T 4 (see Figure 4.34).

T-stop numbers are the same as f-stop numbers but a t-stop is not the same as an f-stop. As I mentioned earlier, f-stops are mathematical calculations based on the size of the diaphragm opening. A t-stop is an actual measurement of the light that is transmitted by the lens at a given diaphragm opening or aperture setting. Many times a lens will be calibrated for both f-stops and t-stops. When setting the exposure precisely on the lens, you should always use t-stops. When measuring the intensity of the light with a light meter or when calculating depth of field, you should always use f-stops.

When the D.P. tells you the f-stop or t-stop to be set on the lens, she may say it in a number of different ways. She may say it is halfway between 2.8 and 4, or the stop is $4^1/_3$, or it is a 3.4, and so on. It is a good idea to work out with the D.P. how she describes the stops so that you understand exactly what she means. See Table E.1 in Appendix E for the intermediate f-stop or t-stop settings for one fourth, one third, one half, two thirds, and three fourths of the way between full stops.

Whenever you film at a frame rate other than 24 f.p.s., you must change the stop to compensate for the new frame rate. If you film at speeds faster than 24 f.p.s., less light strikes each frame of film, so you must increase your exposure. If you film at speeds slower than 24 f.p.s., more light strikes each frame of film, so you must decrease your exposure. Table E.4 in Appendix E shows the f-stop compensation for various changes in frames per second.

It also may be necessary to adjust your exposure when you are using certain filters on the camera. Some filters decrease the amount of light passing through the lens, while others have no effect on the light. Any exposure change will always be an increase, requiring you to open up the aperture. Tables E.2 and E.3 in Appendix E lists some of the most commonly used filters and the amount of f-stop or t-stop compensation, if any, for each. There are many other filters in use that require some type of exposure compensation. Check with the camera rental house about the filters you are using.

It is also necessary to adjust your exposure when you are filming with a different shutter angle set on the camera. Table E.5 in Appendix E shows the f-stop or t-stop compensation for changes in shutter angle.

Changing Lenses and Filters

Whenever you are asked to place a new lens, filter, or any other accessory on the camera, do it as quickly as possible so that the D.P. or Camera Operator can set up the shot. The standard procedure for changing lenses or filters on the camera is as follows. The 1st A.C. tells the 2nd A.C. what the new lens or filter is. While the 2nd A.C. obtains the new item from the equipment case, the 1st A.C. removes the old item from the camera and prepares the camera to accept the new item. The 2nd A.C. brings the new item to the camera and exchanges it for the old item with the 1st A.C. Remember, when exchanging items both assistants should give some type of indication that they have a firm grip on the item so that the other person knows that it is all right to release it. I usually say "got it" when exchanging items with my assistant. This lets him know that I have a firm grip on it and he can let go. While the 1st A.C. places the new item on the camera, the 2nd A.C. places the old one back in the equipment case. The 1st A.C. makes the camera ready for any other accessories, while the 2nd A.C. obtains the accessory from the proper equipment case.

As stated in Chapter 3, don't leave an equipment case alone without closing the case and securing at least one of the latches on the case. There have been a few times when I have picked up a case that my assistant forgot to latch. Fortunately I realized it in time before any of the contents spilled out. During filming, there are many different camera setups, and the equipment must be moved many times during the day. If a case is not latched and someone else picks it up to move it, there could be disastrous results.

Before placing any lens or filter on the camera, check it for dirt, dust, or scratches. If the lens or filter requires cleaning, first place it on the camera for the D.P. or Camera Operator to approve. Once it has been approved, it may be removed and cleaned before shooting the shot. When a new lens is placed on the camera, set the aperture to its widest opening and the focus to an approximate distance to the subject. Remember to engage the follow focus gear and adjust the position of the matte box if necessary. Make sure there is no *vignetting*, which means that you should not be able to see the matte box or lens shade when looking through the lens. Look through the eyepiece after changing a lens to be sure that it is focused properly, and to check that the shutter is cleared for the Camera Operator to view the scene. If you are not able to look through the eyepiece, ask the Camera Operator to check for you. Also, when changing lenses you may have to change the lens support rods because of the physical size of the lens. When bringing the lens from the case, the 2nd A.C. should remember to bring the appropriate lens support rods and support brackets when required. Often, when changing lenses it may also be necessary to rebalance the camera, such as when you change from a prime lens to a zoom lens or vice versa.

Remember to check the balance whenever any new piece of equipment has been added to or taken away from the camera. The camera must be balanced properly for the Camera Operator to do her job correctly.

Also, if you are using any filters, a small identification label should be placed on the side of the matte box or camera stating which filter is in use. Without this reminder tag, the D.P., Camera Operator, or 1st A.C. may forget which filter is in place and then forget to compensate the exposure. By placing a tag on the matte box or camera, it reminds the 1st A.C. and the D.P. that there is a filter in front of the lens (Figures 4.35 and 4.36).

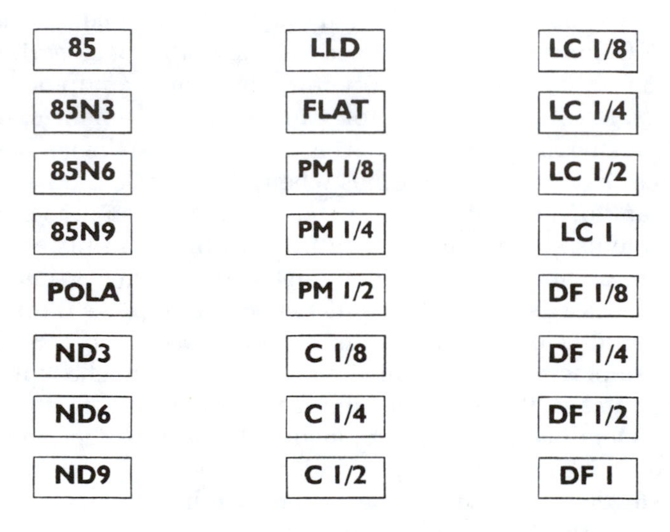

Figure 4.35 Examples of filter identification tags.

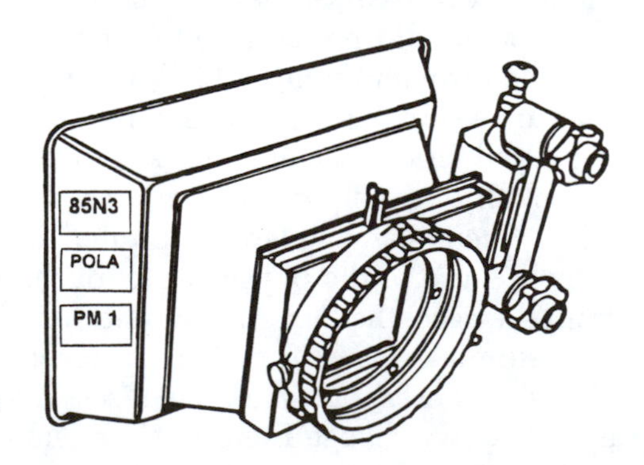

Figure 4.36
Filter identification tags in place on matte box.

Focus Measurements and Following Focus

During rehearsals the 1st A.C. measures the distance from the camera film plane to the subject for each subject position and each camera position of the shot. For beginners, it is important to remember that the focus measurement is taken from the film plane of the camera, to the actor or subject. The film plane is the point in the camera where the image comes into focus on the film, and it is from this point that all focus measurements are taken. If it is not possible or convenient to measure to the actor during rehearsals, obtain focus marks by measuring to the positions of the stand-ins. Just before you get ready to shoot the scene, double-check these focus measurements when the actors step in. This is especially important on scenes that involve critical focus marks where you have very little depth of field. After a shot has been completed, if you have any doubts about the focus, ask for a moment to check the focus of the actor on his mark to determine if focus was good or not. When obtaining the focus measurements, try to do it as quickly and unobtrusively as possible without interfering with the Director, actors, or other crew members. I have heard it said that a good camera assistant is one who is quick, invisible, and quiet. There are so many people on the set that any idle chatter or unnecessary noise tends to be distracting to crew members trying to work and also to the actors trying to rehearse their lines. It is important to remember to never let anyone rush you when obtaining your focus marks or distances. The most beautiful lighting, set design, makeup, and performance is not worth anything if the shot is out of focus.

When obtaining your focus mark or measuring the distance to subjects, there is a special situation that often arises that you must be aware of. This situation is when you are filming the reflection of a subject in a mirror. When shooting a reflection, you must measure the distance from the camera to the mirror and then to the subject. For example, if the distance from the camera to the mirror is ten (10) feet, and the distance from the mirror to the subject is five (5) feet, then you would set the focus of the lens to fifteen (15) feet (10 + 5 = 15), in order to have the reflection of the subject sharp and in focus.

If an actor and camera are stationary, focusing is not very difficult. When an actor or camera or both are moving, focusing during the shot becomes more challenging and even more fun. When an actor is moving in the scene, such as walking toward or away from the camera, the 1st A.C. places tape marks or chalk marks on the ground as reference points for focusing. These focus marks are usually placed about 1 foot apart, or placed according to the markings on the lens. When I am getting focus marks, I base my marks on how the lens is marked. If the lens has focus marks at 20, 15, 12, 10, 8, 7, 6, and 5 feet, I place focus marks according to these distances. When following focus it is much easier for me to hit an exact mark if it is already marked on the lens rather than having to guess. Because of the principles of depth of field, focus marks are not as critical when using a wide lens, and you do not need to measure to as many points as you would if you were shooting with a long focal length or telephoto lens. As an actor passes these marks, the 1st A.C. adjusts the focus to correspond to the distance measured to each point. If the ground is seen in the shot, the 1st A.C. measures to various places on the set, such as pieces of furniture; paintings on the wall; trees, shrubs, or rocks if you are filming outside; and so on. If the camera is mounted on a dolly and will be moving during the scene, the assistant places marks at 1-foot intervals or according to the marks on the lens. As the dolly moves past these marks, the assistant adjusts focus to correspond with each mark. For each distance measured, the 1st A.C. marks the lens or focus-marking disk accordingly so that she may rack focus or follow focus during the scene. The focus-marking disk is usually a white piece of plastic that is attached to the follow focus mechanism. Using a grease pencil or erasable marker the 1st A.C. marks the disk according to the distances measured for the shot. Some assistants wrap a thin piece of tape around the barrel of the lens and place their focus marks on it for the shot (Figures 4.37 and 4.38).

In addition, the assistant may place a reminder tape near the lens with the distances marked on it for the particular shot. It is usually a good idea for the 1st A.C. to keep a small note pad with her to record the focus distances and lens sizes for each scene. Many times you may do a shot of one actor for a scene, and then later in the day you need

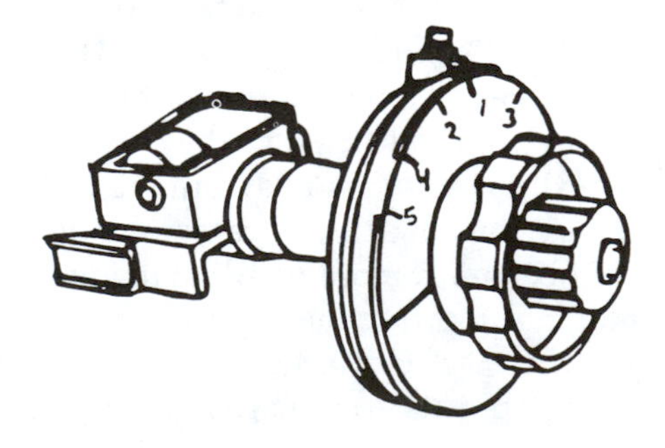

Figure 4.37
Focus marking disk on follow focus mechanism marked for following focus. (Courtesy of Panavision, Inc.)

Figure 4.38
Lens marked for following focus. (Courtesy of Panavision, Inc.)

to do a similar shot of another actor or actors for the same scene. The shots should be made with the same focal length lens and at the same distance as the first shot so they will match when edited together. If you have the numbers written down for the previous shot, it will be no problem to match the focal length and distance for the other shots.

In addition to the 1st A.C.'s focus marks mentioned earlier, the 2nd A.C. places marks for each of the actors' positions during the scene. These marks are not only for the actors to know where to stand during the shot, but also for the 1st A.C. to use for focusing purposes. There are times when actors don't stop on their marks, so you have to estimate their distance from the camera. When you are filming on a sound stage, permanent sets, or practical locations, you often can measure the length and width of each set and record these distances on a sheet of paper for future use. This way if you are in a rush situation and are unable to obtain all of your focus measurements, you can estimate the distance based on where the camera is placed on the set. After a while you should become experienced at guessing the distances with some degree of accuracy. If you have a complicated focus move to do, request at least one rehearsal before shooting the scene.

You also can obtain the focus marks by looking through the eyepiece and focusing on the subject by eye. You then make a mark on the lens to indicate the focus. Always open the aperture on the lens to its widest opening when obtaining focus marks by eye. On a zoom lens you should zoom in to the tightest focal length to obtain an eye focus. Once you have the focus mark, return the zoom to its correct focal length for shooting. On all lenses, remember to set the correct t-stop setting after obtaining eye focus marks.

Following focus or pulling focus is a very precise and exact job, and can only be learned by actually doing it. It takes much practice and experience to be able to do it well and cannot be explained fully in any book. One important thing to remember when pulling focus is to keep a very light touch on the follow focus mechanism. The Camera Operator must follow the action within a scene, and does not want anything to prevent smooth pans or tilts with the camera because the 1st A.C. had a tight grip on the focus knob.

Zoom Lens Moves

The 1st A.C. may be required to do a zoom lens move, which means that the assistant will be changing the focal length of the zoom lens during the shot. The focal length of the lens may change from wide to tight, or from tight to wide or anywhere in between. The important thing to remember when doing a zoom lens move is to start and end the zoom move very smoothly. Any sudden starts and stops are distracting when they are viewed on the screen. The principle for starting or stopping a zoom move is similar to the way you take off in your car or stop your car at a traffic light. Most people start out slowly and work up to the proper speed. The same thing applies to the beginning of the zoom lens move. Start the zoom move slowly and work up to the proper speed so that it does not look like a jerky, quick start. As you start to reach the end of the zoom move, you should slow the speed down until you stop completely.

Many zoom lens changes are done along with some type of camera move, either panning, tilting, dollying, or booming. When doing any type of zoom lens change along with a camera move, the zoom lens change should start a fraction of a second after the camera move starts and end a fraction of a second before the camera move ends. This helps to hide any sudden starts or stops in the zoom lens move and makes it less noticeable to the viewer.

Most zoom controls and zoom motors have some type of switch that allows you to adjust the speed of the zoom. During the rehearsal work out the correct zoom speed with the Camera Operator. If you have a complicated zoom move to do, request at least one rehearsal before attempting to shoot the scene.

There may be some instances when you have to do a zoom lens move for a shot on a lens that does not have a zoom motor. The main thing to remember in this case is to keep a light touch on the lens so you are not restricting the Camera Operator's movements. Just like following focus, zoom lens moves require much practice and experience to be able to do them well and cannot be explained fully in any book.

Footage Readings

After each take, the 1st A.C. should call out the dial readings from the camera footage counter to the 2nd A.C. These amounts are entered in the correct space on the camera report for the particular shot. To make the mathematics easier when totaling up the figures on the camera report, round all dial readings to the nearest ten. As you probably learned in elementary school, if the number ends in four or less you round down, and if it ends in five or more you round up.

Example: The camera footage counter shows a reading of 274. Because this number ends in a four, we round down, and it becomes 270. For this dial reading the 1st A.C. will drop the zero at the end and call out "twenty-seven" (27). The 2nd A.C. will then record "270" on the camera report for that particular shot.

Most often after a take the set becomes very noisy. The Director may be talking to the actors, the D.P. may be giving instructions to other crew members, and so on. It is not a good idea to add to the noise by calling out the dial readings. There is a standard set of hand signals used to give dial readings. They also can be used when the 2nd A.C. is too far away from the 1st A.C., so she does not have to shout across the set. Figure 4.39 shows the standard hand signals used for footage counter dial readings.

Checking the Gate

After each printed take, it is standard procedure for the 1st A.C. to check the gate for hairs, which are usually fine pieces of emulsion or dust that may have gotten in the gate and will show up on the screen as a large rope in the frame. When I say *check the gate*, I am referring to the opening in the aperture plate that the light passes through from the lens to the film. Most often if a hair is found, you should do another take to be sure that you have a clean shot. The D.P. will usually look at the gate to double-check because he has the final say on whether you should do another take. Many times another take is not necessary even if there is a hair. The hair may be so small that it does not reach into the frame.

There are three accepted ways to check the gate for hairs: remove the gate, remove the lens, and look through the lens.

1. Remove the gate: Turn the inching knob so that the registration pin and pull-down claw are away from the film. Remove the film

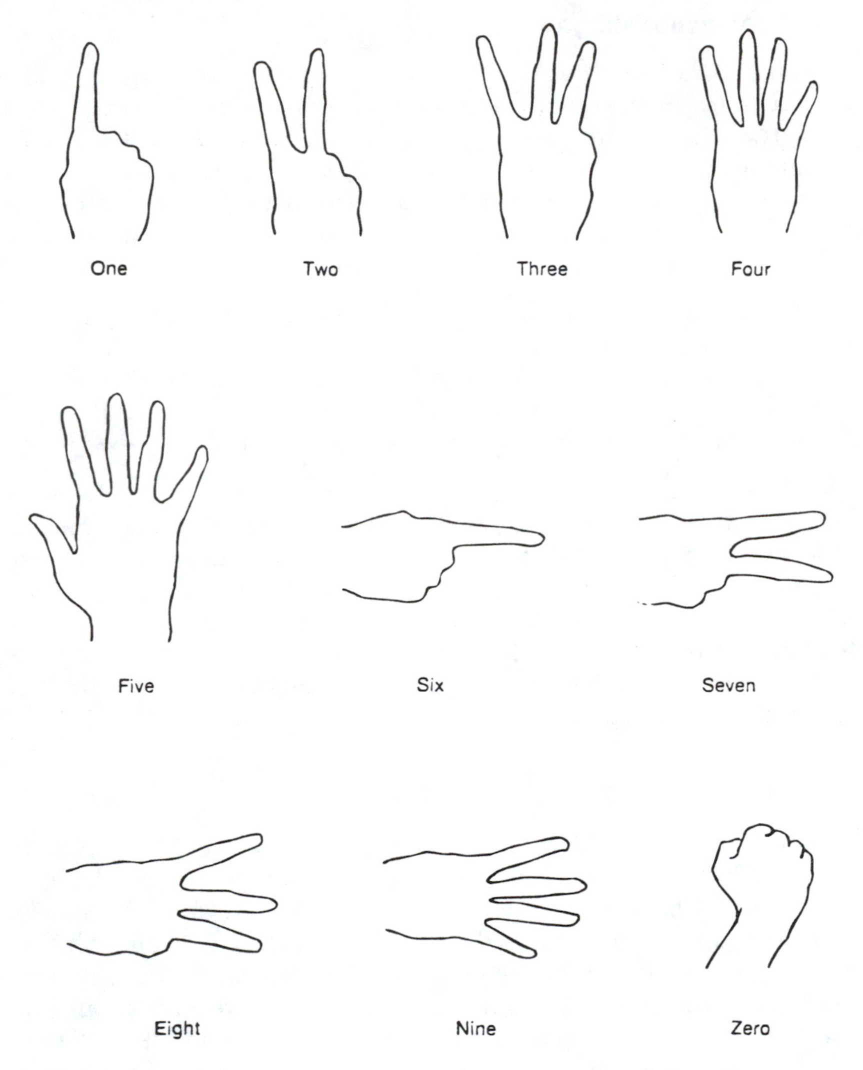

Figure 4.39 Hand signals for camera footage counter dial readings.

from the gate and then remove the gate. When you hold the gate up to the light, you should be able to see any hair along the edges. This is not always an accurate way to check for hairs because when you remove the film from the gate, the hair may stick to the film and be pulled out with it. When you look at the gate, you will not see any hair and assume that the gate is clean. You are not able to remove the gate on many cameras, so this method may not be used on all cameras.

2. Remove the lens: Remove the lens from its mount. Then turn the inching knob to advance the shutter, until you can see the emulsion in the gate while looking through the lens port opening. Using a small flashlight or magnifier with a built-in light, examine the emulsion along the top, bottom, and sides to see if there is any hair, which would be visible against the bright background of the film emulsion. This method is the one that I prefer to use whenever possible. If removing the lens is too difficult or time consuming, I will use one of the other methods.

3. Look through the lens: Open the aperture on the lens to its widest opening. Turn the inching knob until the emulsion is visible in the gate as you look through the lens. Place a small flashlight alongside your face and look right down the barrel of the lens. The lens will act as a magnifier for the gate, allowing you to see any hair along the top, bottom, or sides of the gate. This method only works with lenses that are 40mm or longer in focal length. When using this method with a zoom lens, be sure to zoom the lens in to its longest focal length.

If a hair is found, clean the gate and aperture plate with compressed air and an orangewood stick. When cleaning any emulsion from the gate, use only an orangewood stick. Once you have cleaned the gate, double-check it before shooting any additional shots.

Moving the Camera

The camera must be moved frequently throughout the day. If the camera is mounted to a dolly, it is much easier for the dolly grip to wheel it to each new setup. One of the camera assistants, usually the 1st A.C., walks alongside the dolly, with one hand on the camera to steady it while the dolly is moving. The dolly may have to travel over rough terrain or over lighting cables. If the terrain is too rough, it is a good idea to remove the camera and carry it to the next setup. The bouncing of the dolly can shake loose the elements of the lens or possibly even damage the camera. When the camera is mounted on a tripod, it is the sole responsibility of the camera assistants to move the camera to each new position. Many camera assistants pick up the entire tripod, with the head and camera attached, and carry it to the new position. One of the best ways to do this is as follows: Aim the lens so it is in line with one of the legs of the tripod. Lock the pan and tilt locks on the head. Lengthen the front leg of the tripod (the one that the lens is in line with). Crouch down and place the shoulder pad of the tripod between your shoulder and the head with the extended tripod leg in front of you. With your left and right hands grab the left and right legs of the tripod and slowly push them in toward the front leg. These two legs will fold up, forcing all of the weight onto the one extended leg. The camera will lean into your shoulder making it easy to pick up. Stand up, and the camera, head, and tripod should be balanced on your shoulder. To place the entire system back on the ground first crouch down and set the long tripod leg on the ground. Grab the left and right legs, and bring them back to their normal position to form a triangle with the extended leg. Loosen the extended leg and return it to its original length. The camera is now ready for shooting at the new setup. Another way to move the camera on a tripod is for the 1st A.C. to remove the camera from the head and carry it while the 2nd A.C. carries the tripod and head.

Use whichever method is easier and safer for you. Never try to carry anything if you do not feel it is safe or you don't think that you can handle it. Before shooting be sure that the camera is level, whether it is mounted on a dolly or on a tripod. Each time the camera is moved, bring along all other needed equipment and accessories, including lenses, filters, magazines, and so on. When the D.P. requests a piece of equipment, he will not want to wait because you left some of the cases back at the last camera position. If the cases are on a dolly or hand cart, all you need to do is wheel it to the new position. Otherwise the cases must be hand carried to the next setup. It is the camera assistants' responsibility to make sure that the camera equipment is moved quickly and safely, and is near the camera throughout the day. If you require any help in moving or carrying the equipment, do not hesitate to ask one of the production assistants or grips on the set. It is much better to ask for help than to try to do it all yourself and risk getting hurt or dropping and damaging the equipment.

Performing the Duties of Second Camera Assistant

There will be times when you are working on a production that does not have a 2nd A.C. It may be a small production, such as a music video or commercial, or perhaps they just do not have a big enough budget to afford an additional assistant. If this is the case, the 1st A.C.

also carries out the duties of 2nd A.C. Because you are now doing two separate jobs, it is important to remember not to be rushed while working. The Director and D.P. should understand that it sometimes takes a little longer to get certain things done. If for any reason you need help, do not hesitate to ask other crew members.

Packing Equipment

At the end of each shooting day, pack all camera equipment away in its appropriate case and store it in a safe place until the next shooting day. If you are working out of a camera truck, place and secure all of the cases on their appropriate shelves. If you are working on a sound stage, you should have a room or special area where all of the equipment is stored at the end of each shooting day.

Tools and Accessories

As mentioned in Chapter 3, with many professions you must have some basic tools and accessories so that you may do the job properly. When first starting out, you should have a very basic tool kit or ditty bag, and as you gain more experience and work more frequently, you can add things as you feel they are needed. Some of the tools are common tools that you may need, while others are specialized pieces of equipment that are unique to the film industry. In addition to the basic tools, you should also have a small inventory of expendables, film cans, cores, camera reports, etc. There will be many times when you are called for a job at the last minute and you may have no time to acquire some of these items. By having a small amount on hand, you will always be prepared for most job calls that you get.

Many 1st A.C.s will wear some type of belt pouch or fanny pack to keep the most commonly used tools or accessories with them at all times. Instead of wearing a pouch, which can become very cumbersome when packed full of tools and accessories, many first assistants have a box called a *front box*, which contains all of the basic items needed each day for shooting. The front box is most often constructed of wood and has a metal bracket on the back that allows you to mount it directly on the front of the head, under the camera. It will contain items such as a depth-of-field calculator, cloth and metal tape measures, sharpies, miniflashlight, slate markers, lens tissue, and lens cleaner, etc. The front box may also be used to hold the D.P.'s light meters. By mounting this on the head, the 1st A.C. has the basic supplies needed for shooting and does not have to be encumbered by wearing a large pouch filled with these tools and supplies. An illustration of a front box is shown in Figure 4.40.

See Appendix D for a list of the common tools and equipment that should be included in an Assistant Cameraman's ditty bag or tool kit.

POST-PRODUCTION

Wrapping Equipment

At the completion of filming, the camera equipment, camera truck, and anything else relating to the camera department must be wrapped. This means that everything should be cleaned and packed away. All equipment must be cleaned, packed, and sent back to the camera rental house. The wrap can take anywhere from a few hours to an entire day, depending on the size of the camera package. Usually the 1st A.C. wraps out the camera equipment, while the 2nd A.C. wraps out the truck, darkroom, and film stock. Many times, if it is a small production, only the 1st A.C. does the wrap. This process usually takes a few hours, or possibly a whole day, and is usually done the same day shooting ends or the day after shooting has stopped. Clean all camera equipment and place it in the proper case. Remove any identification labels placed on the equipment during the camera prep before putting the equipment in the case. The cleaning of all cases and equipment may seem like a lot of wasted work, but it lets the rental house know that you are a professional and care about the

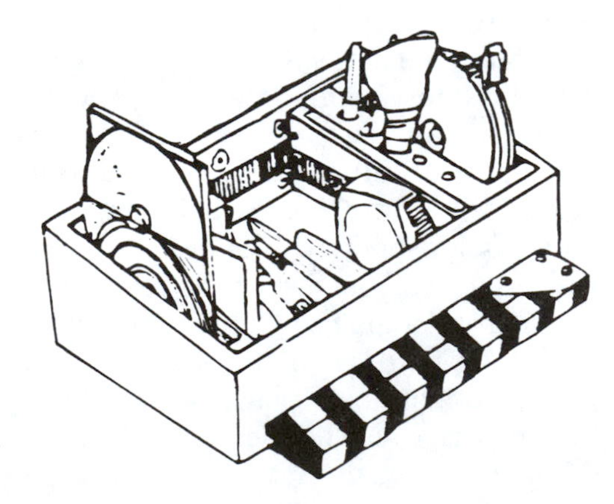

Figure 4.40
Illustration of front box used by most 1st A.C.s.

equipment you work with. The rental house will also feel that the next time they send out equipment for you they do not have to worry about the equipment. You should have copies of all packing lists for all equipment received since the beginning of the production. If you find anything missing, notify the production office immediately so that they are not surprised when the rental house calls them.

5

Problems and Troubleshooting

Troubleshooting may be described as a careful system of finding the cause of a problem and correcting it. When something goes wrong find out why, and then correct or eliminate the problem. You need common sense and logic, and also a knowledge of the equipment you are working with. If you are familiar with the equipment you should be able to follow a step-by-step procedure to find and correct almost any problem that you encounter.

Being familiar with the equipment not only involves the ability to put the pieces together, but also feeling comfortable with the equipment. Treat the camera and its accessories gently. Do not force any pieces of equipment that will not fit together. When placing the camera on the head or base plate, slide it on gently and do not just slam it in place. The better you take care of the equipment, the fewer problems you should have.

When you do encounter a problem, the first and most important thing is not to panic. Think about what the problem is and then decide what is the most logical cause. Try to fix it, and if it doesn't work, continue trying to correct the problem by process of elimination. Try the obvious first, eliminate what is not causing the problem, and eventually narrow down the possible choices and find out what the cause is.

It is important that you check only one thing at a time. For example, if the camera won't run, and you change the battery, the power cable, and the fuse at the same time, how will you know which of these was the cause of the problem? Finding the cause of most problems should usually take only a few minutes, but there will be some instances when you cannot find the cause yourself and then must telephone the rental house and ask for help. Never be afraid to contact the rental house regarding any questions you may have about the equipment. They would rather have you ask about something than to try to do it incorrectly.

Because it is not possible to foretell every problem, I mention some of the typical ones that can and will happen in the course of a film production. Many of these things have actually happened to me on various shoots. Sometimes I was able to correct the problem quickly, but other times I had to call the rental house and ask for their advice, or have a technician come to the location to fix or replace the camera or accessory.

Problem: Battery loses power.
This is one of the most common problems you will encounter. The battery may not be fully charged or may be unable to hold a charge. Try to completely discharge the battery and then place it on charge overnight. If this does not solve the problem, then have the battery checked and replaced if necessary. You should never go out on a shoot with only one battery, just in case this does happen. By having an additional battery, you will not have to stop filming until another one can be obtained. At the very minimum you should have at least two batteries.

Problem: Lens will not focus.
One reason could be that the lens is not seated properly in the lens mount. This means that the lens is not correctly mounted on the camera. Remove the lens and check the lens mount of the camera to see if there are any obstructions. Also check the back of the lens to see if there is anything that would prevent the lens from mounting properly. If the lens and the lens mount seem all right, re-insert the lens and check the focus again.

Another reason that the lens might not focus is that the ground glass of the camera is inserted incorrectly. Check the ground glass and re-insert it if necessary. The ground glass should be inserted into the camera with the matte or dull side facing toward the mirror of the shutter.

Another cause of lens focus problems is that the lens is damaged. Check this by using a focus chart and checking the lens as you originally did during the camera prep. Place the chart at various distances from the camera, and then check the eye focus mark on the lens to see if they match. If a problem is detected when comparing the eye focus to the measured focus, the lens should be returned to the rental house for repair and a replacement lens should be obtained.

One of the most common reasons that the lens appears out of focus when looking through the eyepiece is that the diopter adjustment of the eyepiece was changed when you were away from the camera. Each time anyone looks through the eyepiece, check that the diopter adjustment ring is set to the appropriate mark. Because each person's vision is different, the image may look sharp and in focus to one person, but blurry and out of focus to another.

Problem: Camera will not run.
When this happens, first check to see if the battery is connected to the camera. This is the most logical and most common reason why the camera won't run. Also check to see if the battery contains a full charge. If the camera has one, check the buckle trip switch inside the camera to see if it is in the proper position. Reset the switch if necessary, which should correct the problem. If the battery is connected and the buckle trip switch is in its proper position, try a new battery cable. If this doesn't work, try a new battery.

Some cameras have a safety feature built into them that will not allow the camera to run if the camera body door is not closed completely. Make sure the door is closed and latched before turning on the camera. If the camera still will not run, change the fuse if you are able to gain access to it. You can change the electronic circuit boards in Panavision cameras. I believe that it is a good idea to first check the fuse before resorting to changing the electronic circuit boards. If all of the above actions still do not correct the problem, call the rental house for help. It is important to remember when this or any problem occurs to check only one thing at a time. Before changing fuses or circuit boards, always disconnect power to the camera.

When trying to determine why the camera will not run, disconnect all electrical accessories from the camera and try to run it. This helps to determine if any accessories are causing the problem. Check all electrical accessories one at a time to see which one, if any, may be causing the problem.

Some cameras require a minimum and maximum voltage amount in order for them to run. Be sure that the battery is functioning properly so that these amounts are within the guidelines for the camera system you are using. Finally, if the camera does not run, be sure to check that the run switch is in the "on" position.

Problem: Camera does not stop when switched to "off."
This can happen often, especially when you are using a hand grip with an on/off switch or a zoom control with an on/off switch. If there

are any accessories plugged into the camera that contain an on/off switch, check to see that this switch is in the off position. If it is in the on position the camera will continue to run when you turn the main power switch off.

When mounting the camera to the head or sliding base plate, you must screw a 3/8 inch-16 bolt into the bottom of the camera. This bolt could be making contact with the camera motor and shorting out the on/off feature, causing the camera to continue to run. Remove the bolt and use a shorter one that does not touch the motor.

Problem: Camera starts and stops intermittently.

The battery might not be fully charged. Changing batteries should correct the problem. If the battery cable is loose, re-insert it into either the camera or the battery. A loose wire in the power cable can also be one of the causes that you might not be able to see by just looking at the cable. If you suspect this, try wiggling the cable at the point where it is connected to the camera and also where it is connected to the battery to see if this causes any change. If the camera starts and stops, it is a good indication that there is a short in it. Try a new cable and have the other one checked as soon as possible. On Panavision cameras you may find it necessary to change the circuit boards to correct this problem. The important thing to remember when changing circuit boards in Panavision cameras is to always change all of the circuit boards at the same time. Never replace just one or two of the boards. Send the old boards back to the rental house and request a replacement set to have on hand.

Problem: Camera is noisy.

The most common reason for this is that the film is not threaded properly in the camera. The top or bottom loop may be too large or too small, causing the movement to work harder moving the film through the camera, which results in the movement being a little noisier than usual. Rethread the camera and set the loops to the proper length. On many cameras the loop may be set when threading the magazine, so you have to either rethread the magazine or change magazines. Check that all rollers are closed and the film perforations are engaged on the sprockets correctly. If the camera has an adjustable pitch control, adjust it so that the camera is running as quietly as possible when it is threaded correctly. Sometimes none of these solutions make the camera any quieter, so the only thing to do is cover the camera with a sound barney to cut down on the noise.

I have also found that some film stocks cause the camera to run noisier than others. Because of differences in the manufacturing of the different film stocks and emulsions, some film may be slightly thicker or thinner than other film. I was once working on a small production and the camera ran noisier than usual. The Director asked me if there was anything that could be done, and I explained that it was because of the film stock we were using. It was a little thinner than the stock we had used previously. The Director was not satisfied with my explanation so he telephoned the person from whom he had rented the camera and asked that he come to the location to check the camera. When he arrived, he asked me what film stock we were using. When I told him what it was, he told the Director that the film stock was causing the noise problem and that there was nothing he could do about it.

Problem: Magazine is noisy.
This may be caused by the film spooling off the core and rubbing against the side of the magazine. Hold the magazine flat in your hands and give the cover of the magazine a firm slap so that the film settles back onto the spool or core. When the magazine is already placed on the camera, give both sides of the magazine a firm slap to force the film back onto the spool. If the film loop is set when threading the magazine, a loop of incorrect size will also cause the magazine to be noisy. To correct this rethread the magazine so that the loops are the proper size. There are some newer magazines that have a timing adjustment. Check with the rental house if you are not sure how to adjust the timing on the magazines that have this feature.

Problem: Camera door does not close.
On some cameras the door does not close when the sprocket guide rollers are not closed, or when the movement is pulled away from the gate for threading. Be sure that both of these are in the correct position for filming, and the door should close properly. Also check the edges around the door where it meets the camera to be sure that there are no obstructions. Clean out anything that may be blocking the door and it should close properly.

Problem: You are unable to thread film into the gate area.
Be sure that you have turned the inching knob to advance the pull-down claw so that it is withdrawn from the aperture plate. Also, check to see if the registration pin is withdrawn from the aperture plate. These are the two most common reasons why you cannot thread the

film. If after checking these two items you still encounter the problem, check to see that there are no film chips stuck in the gate area, preventing the film from threading properly.

Problem: Film does not take up.
If you are using an older camera that uses some type of belt to drive the take-up side, check that you have the right size belt and that it is connected properly to the magazine. You should always have a spare drive belt with the camera equipment. On magazines using the drive belt, the belt must be placed on the correct side of the magazine for it to take up properly. It should be either connected to the feed side or the take-up side, depending on whether the film is going forward or backward. Check the belt and adjust it as necessary. Check with the rental house so that you are sure how to connect the belt properly. Panavision magazines have electrical contacts built into them so that when connected to the camera, the torque motor of the magazine receives power. If these contacts are dirty, the film will not take up properly. Check the contacts and clean them if necessary.

Some cameras also have the ability to run either forward or in reverse. If the switch is in one position on the camera and the opposite position on the magazine, the film will not take up properly. Make sure that the switches on the camera and magazine are in the same position. Check the take-up side of the magazine to see if the film end has come off of the take up core. Rethread the camera and this should correct the problem.

Problem: Camera stops while filming.
The most common cause of this problem is that the film jams in either the camera or the magazine. Check that the loops are the proper size and adjust them if necessary. Again, if the loop is set in the magazine, it must be removed from the camera and rethreaded or a new magazine placed on the camera. When the film jams it can become ripped or torn, leaving a piece of film stuck in the magazine throat or in the gate of the camera. The important thing to remember when clearing any film jam is not to force any part of the camera or magazine. Gently slide the film from side to side or up and down until it will come out cleanly. If you try to force it out, you can damage the camera movement or the gears of the magazine. After clearing any film jam, clean the camera with compressed air to remove any film chips or emulsion that may have become trapped in the gears of the movement.

Another common cause of the camera stopping is that the film has rolled out. Many cameras have a safety feature built into them that shuts off power to the camera when it runs out of film. Be aware of the footage counter when filming so this does not happen.

Problem: Film jams in camera.
The film loop could be the wrong size. Rethread the camera or the magazine and adjust the loops to the proper size. If the magazine is not threaded properly, it can cause the camera motor to work harder to move the film through the camera, which results in the film becoming jammed.

Problem: Film rips or has torn perforations.
The common cause of this is the same as for film jams. Improper loop sizes can result in the film becoming ripped or torn as it goes through the camera or magazine. Check the loop size in the camera and adjust if necessary. Rethread the magazine and adjust any magazine loops accordingly.

Problem: Film looses loop.
Check the pull-down claw and registration pin to be sure that they are not bent in any way. Incorrectly threading the camera or magazine can cause loss of the loop. When threading, be sure that you set the correct loop size in the camera or magazine. Also check to be sure that the film is properly engaged on the sprocket rollers and that the sprocket roller guides are engaged correctly.

Problem: There are scratches on the film.
Whenever the film scratches, scratch test the entire system exactly as you did during the camera prep (see Chapter 4). The cause of the scratches could be from a problem inside the magazine throat or in the gears or rollers of the magazine. It could also be coming from inside the camera at any number of places. There may be dirt or emulsion buildup in the gate that should be cleaned out before you continue to shoot. Dirt or dust in the magazine can also cause scratches on the film. The best way to determine where the scratch is occurring is to place the magazine on the camera and thread the camera normally. Run some film through the camera. Using a permanent ink marker, place an "x" on the film at the following places: where it exits the magazine, enters the gate, exits the gate, and re-enters the magazine. Check the film at these marks to determine where the scratches occurred. An incorrect loop size may also cause scratches on the film. If necessary, you may have to send some of the magazines, or even the camera, back to the rental house for replacement or repair.

Problem: Camera does not run at sync speed.
One cause of this could be a problem with the battery. A weak battery could affect the speed of the camera. Replace the battery with one that is fully charged and the camera should run at sync speed. Another common cause is that the motor switch on the camera is set to the variable position instead of the sync position. Reset the switch to the sync position. On Panavision cameras, a malfunctioning circuit board could also cause this to happen. If you are able, change the circuit boards to see if this corrects the problem. Also, if the magazine or the camera is threaded incorrectly, it may have an affect on the motor, causing it to lose speed. Check the threading of both and adjust as necessary.

Problem: Viewing system is blacked out. (You cannot see anything when looking through the eyepiece.)
This could be one of a number of problems. The shutter may be closed, which makes the eyepiece dark. Turn the camera on and off quickly, or turn the inching knob to clear the shutter. The eyepiece may be set to the closed position, which allows no light to enter the eyepiece. Check the eyepiece control lever, and set it to the open position for viewing. When the lens is stopped down to its smallest opening, it may be difficult to see anything when looking through the eyepiece. Also, if there are any neutral density filters in front of the lens, it darkens the image when viewed, making it appear totally dark. The most obvious reason for not being able to see anything through the eyepiece is that there is someone or something blocking the lens, or possibly that the lens cap is in place. Remove the lens cap or whatever is blocking the front of the lens so that you can see clearly.

When doing certain special effects shots, you may be using a camera that contains a rack-over viewing system. If the viewfinder is not in the correct position, you will not be able to see when looking through it. Be sure to place it in the correct position for viewing.

Problem: Zoom lens motor runs erratically.
There may be a short in the zoom control or in the power cable from the motor to the control. Replace the zoom motor power cable, and if this still doesn't work, replace the zoom control. Check the motor gear where it attaches to the lens. There may be some chips in the gear teeth or lens gear teeth that could cause the motor to slip. Replace the motor gear or lens gear as necessary. The zoom motor may also run without your having to touch the zoom control. Some zoom controls have an adjustment inside the control that must be set

to prevent the motor from running without being engaged. Be sure to check with the rental house before attempting to take apart any piece of equipment.

Problem: Tripod head does not pan or tilt.
The most obvious cause of this is that the pan and tilt locks are engaged. Check the locks for each and release them if necessary. Check the head to be sure that there are no obstructions that could prevent the head from panning or tilting. Remove any obstruction, and the problem should be corrected. Never force the head in either direction. You may worsen the problem, making it impossible for you to correct in the field. The head must then be sent to the rental house for repair and a replacement head sent to you. On gear heads, check that the gear adjustment lever is not in the neutral position. When in neutral, turning either the pan or tilt wheel has no effect on the head. Place the pan and tilt gear adjustment lever in one of the gear positions, which should allow you to pan and tilt with ease. On most gear heads there are usually two sets of locks for the pan and tilt. One is for the pan and tilt controls, to lock them in position, and the other is for the pan and tilt movements, to physically lock them in position, even when the gears are in the neutral position.

Problem: Tripod legs do not slide up and down.
Quite often, when in a hurry you may forget to release the locks for the legs before attempting to adjust the height of them. By releasing the tripod leg locks, they should slide up and down smoothly. Tripod legs get dirty after much use, and usually begin to stick when you try to adjust them. Keep the legs clean on a regular basis and spray them with silicone spray. This should keep them in working order and help them to last longer.

Problem: The image on the video monitor is out of focus or tilted to the side.
On most camera video taps there are adjustments for both the focus and the position of the image on the outside of the video tap. Turn these adjustment knobs until the image comes into focus or the image is in the correct position. If this still does not correct the problem, remove the cover of the video tap, if you are able to, and turn the adjustment knobs located inside. Sometimes if the video tap is not firmly mounted to the film camera, the image appears tilted or out of focus. Check to be sure that it is mounted securely and correctly to the camera.

Problem: When filming in or around salt water, the camera and magazine falls into the water.
Before taking any equipment from the rental house, if you know that you will be filming around salt water, ask them what you should do if any of the equipment falls into the water. The following procedure is the accepted method, but you should check it with the rental house before hand just to be safe. If this should ever happen to you, the first thing that you should do is to completely rinse the camera in fresh water as soon as possible. Don't worry about getting the camera wet. Salt water is highly corrosive and can damage the working parts of the camera very quickly. The faster it is removed, the fewer problems you should have. Don't allow a fully loaded magazine of film to dry. Rinse off the magazine completely, with the film still inside, and ship the entire magazine, packed in water, to the lab for processing.

Problem: A fuse blows when connecting any electrical accessories.
This is a common problem that can be easily corrected. The important thing to remember when connecting any electrical accessories is always to disconnect the power to the camera before attaching the accessories. If the camera is connected to a power source, the connection of any electrical accessory may cause a power surge to the motor, which will then cause a fuse to blow.

Problem: You are shooting outside, using tungsten-balanced film, and do not have an 85 filter available to correct the exposure.
If you are using negative film, the lab can usually correct the color during processing. If necessary you may use an orange color gel, which is the same color as the 85 filter, in front of the lens. Eastman Kodak manufactures gel material that is the same as many of the most common filters. Purchase one of these Kodak gels at most amateur photography stores. They are available for only a few dollars each. Many assistants carry these gels in their ditty bags just in case they encounter this situation.

Problem: You are shooting inside with tungsten light, but only have daylight-balanced film available and no 80A filter for correction.
This is similar to the previous problem regarding tungsten film outside. You may use a gel in front of the lens to correct the exposure. You can also instruct the lab to make the necessary corrections during processing. A few years ago a production company that I was working for mistakenly purchased daylight-balanced film for a shoot that was being done entirely on stage using tungsten lights. I sent a production

assistant to a local camera shop to purchase a Kodak 80A gel filter. I taped the filter to the optical flat in the matte box, and the D.P. made the necessary lighting and exposure changes. We shot the commercial and it turned out just fine.

Try the obvious solution first and then continue in a step-by-step manner until you find out what the cause of the problem is. You will most likely encounter some different problems from those listed here, but if you are familiar with the equipment, you should have no trouble finding and correcting most any problem that you come across. If you are not sure of how to fix a particular problem, call the rental house for their help. Most rental houses will send a technician to your location if the problem cannot be fixed by you in the field.

6

Cameras

As a camera assistant you will be working with many different camera systems throughout your career. You should be familiar with as many different cameras as possible. This section contains basic information such as format, magazine sizes and threading diagrams of cameras, magazines or both for most of the cameras currently used in the film industry.

The threading diagrams included here are not meant to teach you how to load the magazines or thread the cameras. They are only intended as a reference, in case you have forgotten about a certain camera system. If you want to learn how to load magazines or thread cameras, you should contact a camera rental house that has the particular camera you are interested in.

Remember: These illustrations are only to be used as a reference.

Aaton XTR-Prod

Format: 16mm
Magazine sizes: 400′ coaxial

For camera and magazine illustrations and threading diagrams, see Figures 6.1–6.3.

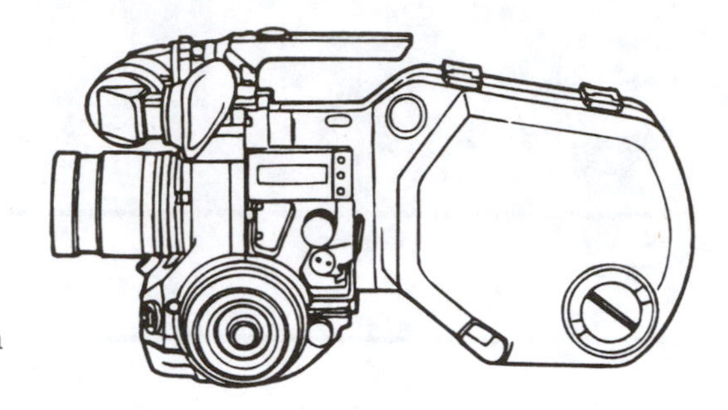

Figure 6.1
Aaton XTR-Prod 16mm camera. (Courtesy of Aaton Des Autres.)

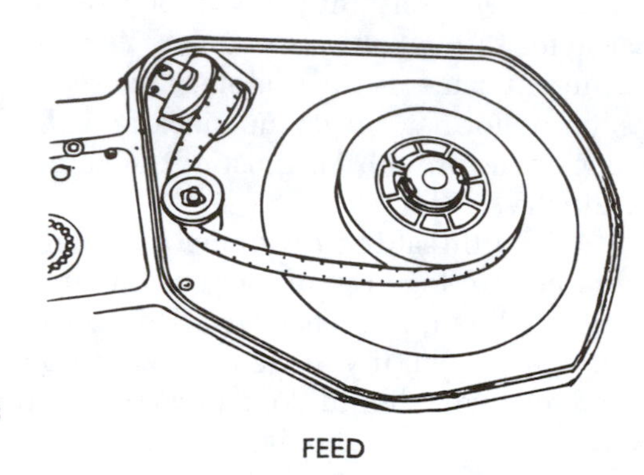

Figure 6.2
Aaton XTR magazine—feed side. (Courtesy of Aaton Des Autres.)

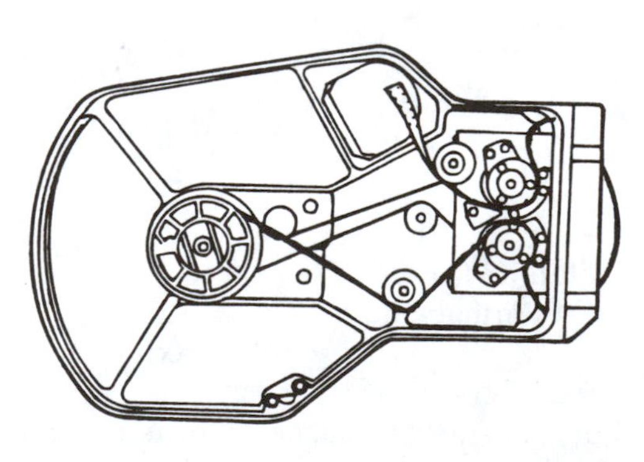

Figure 6.3
Aaton XTR magazine—take-up side. (Courtesy of Aaton Des Autres.)

Aaton 35

Format: 35mm

Magazine sizes: 400′ displacement

For camera and magazine illustrations and threading diagrams, see Figures 6.4 and 6.5.

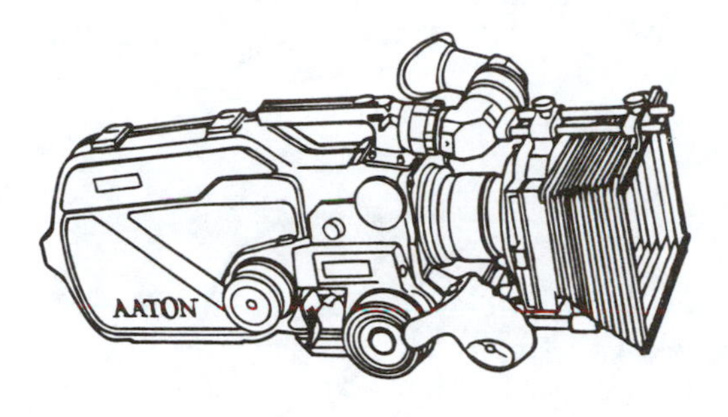

Figure 6.4
Aaton 35 camera. (Courtesy of Aaton Des Autres.)

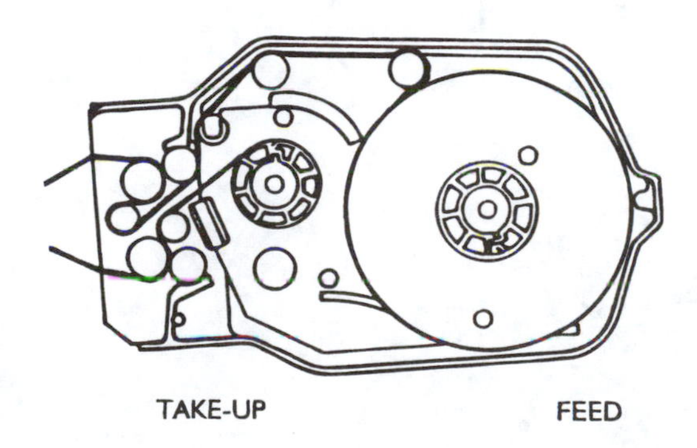

Figure 6.5
Aaton 35 magazine. (Courtesy of Aaton Des Autres.)

Arriflex 16BL

Format: 16mm

Magazine sizes: 200′ and 400′ displacement

For camera and magazine illustrations and threading diagrams, see Figures 6.6–6.9.

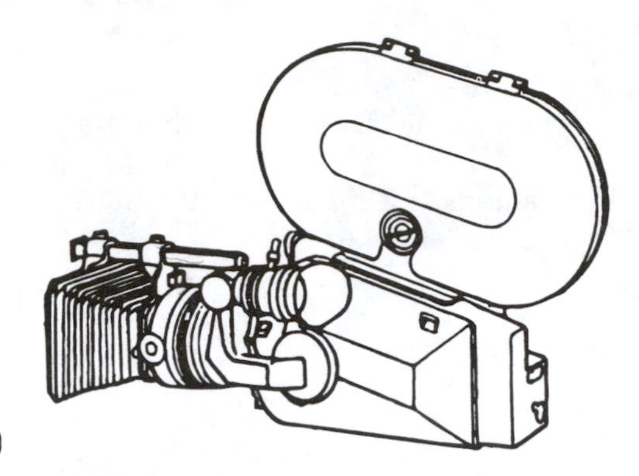

Figure 6.6
Arriflex 16BL camera.
(Courtesy of the Arriflex Corp.)

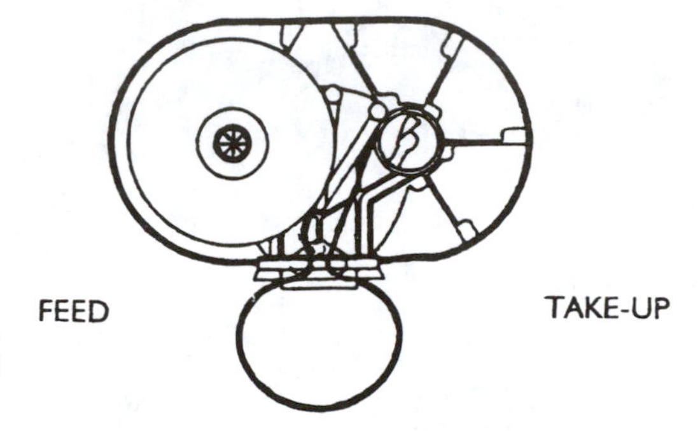

Figure 6.7
Arriflex 16BL magazine.
(Courtesy of the Arriflex Corp.)

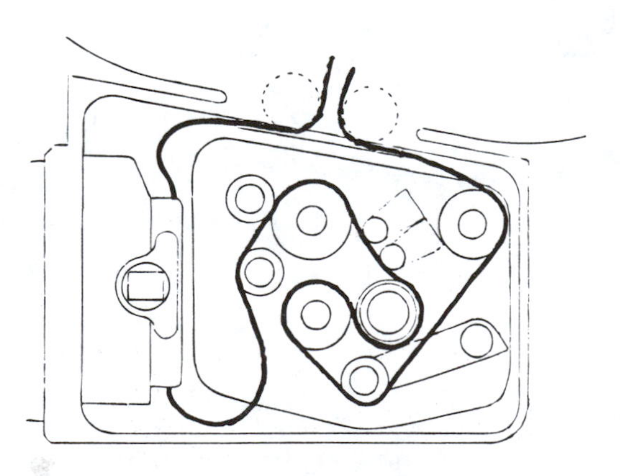

Figure 6.8 Arriflex 16BL camera threading—single system sound. (Courtesy of the Arriflex Corp.)

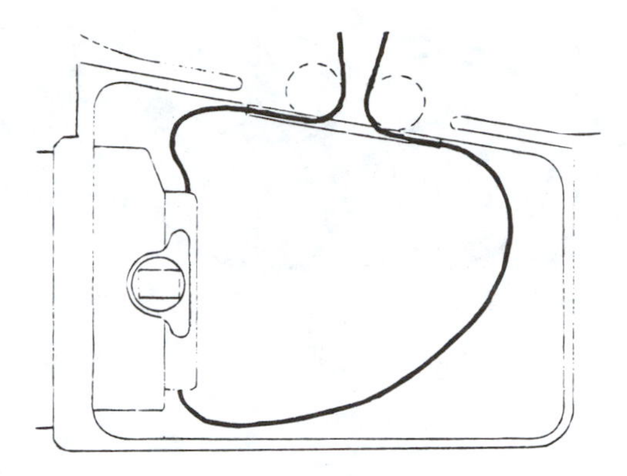

Figure 6.9
Arriflex 16BL Camera threading—double system sound. (Courtesy of the Arriflex Corp.)

Arriflex 16S/SB

Format: 16mm
Magazine sizes: 200′ and 400′ displacement

Note: Camera also has the ability to accept 100′ daylight spool internal load.

For camera and magazine illustrations and threading diagrams, see Figures 6.10–6.12.

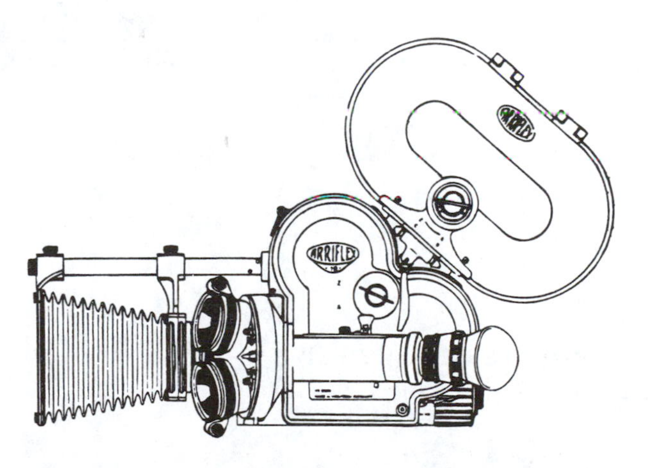

Figure 6.10 Arriflex 16S/SB camera. (Courtesy of the Arriflex Corp.)

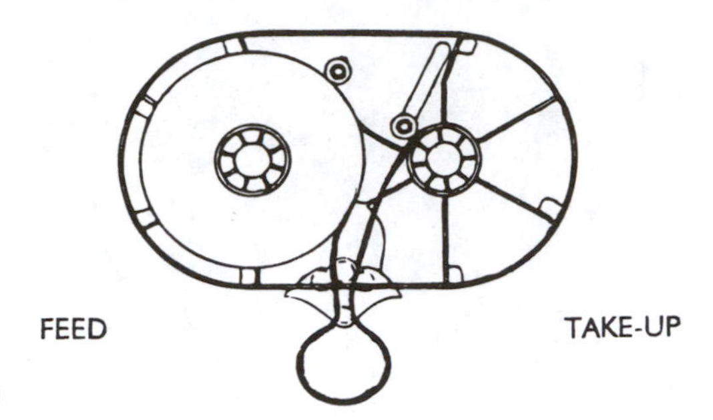

Figure 6.11
Arriflex 16S/SB magazine.
(Courtesy of the Arriflex Corp.)

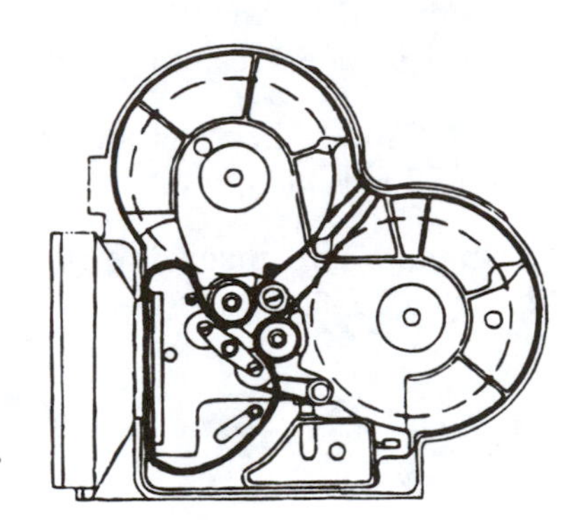

Figure 6.12
Arriflex 16S/SB camera threading.
(Courtesy of the Arriflex Corp.)

Arriflex 16 SR-1, 16 SR-2, and 16 SR-3

Format: 16mm (high-speed model also available)
Magazine sizes: 400′ coaxial

For camera and magazine illustrations and threading diagrams, see Figures 6.13–6.16.

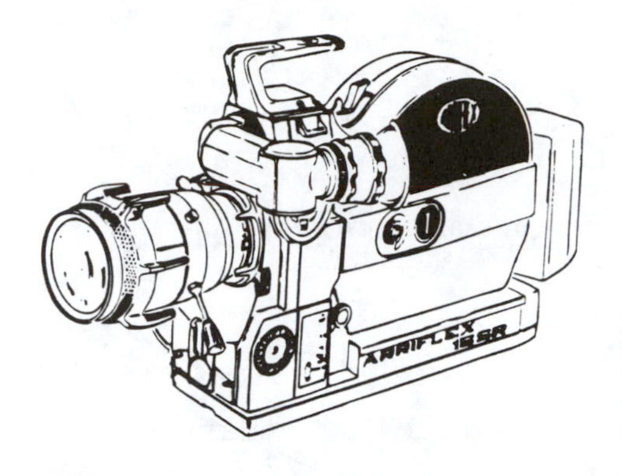

Figure 6.13
Arriflex 16 SR-1 and 16 SR-2 camera. (Courtesy of the Arriflex Corp.)

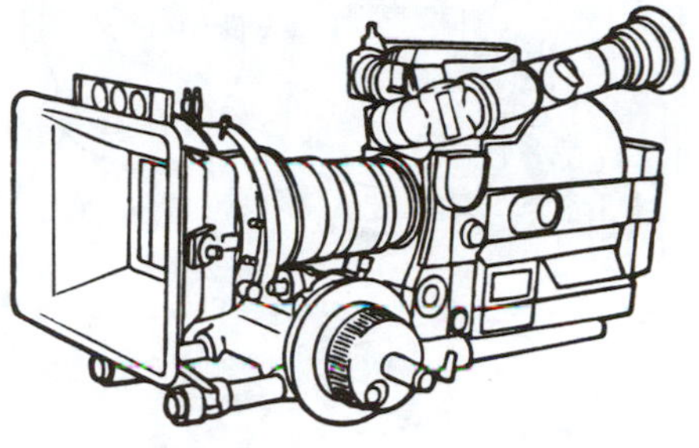

Figure 6.14
Arriflex 16 SR-3 camera. (Courtesy of the Arriflex Corp.)

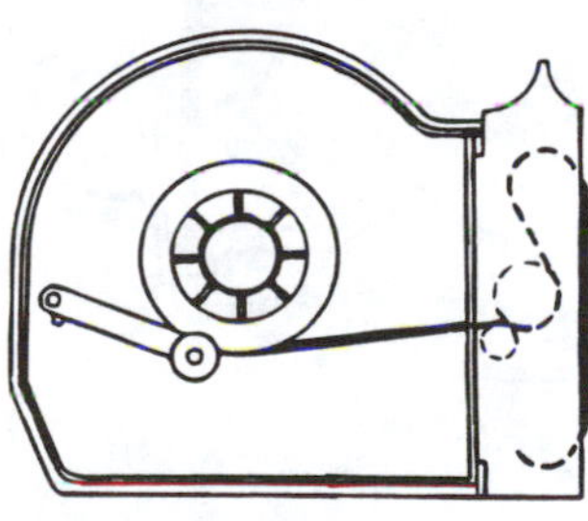

Figure 6.15
Arriflex 16 SR magazine—feed side. (Courtesy of the Arriflex Corp.)

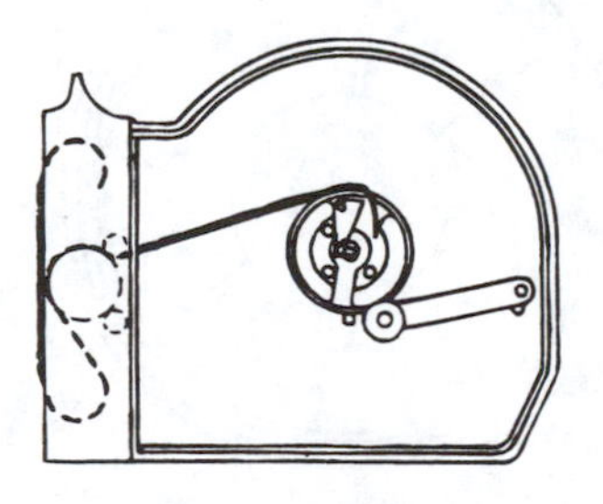

Figure 6.16
Arriflex 16 SR magazine—take-up side. (Courtesy of the Arriflex Corp.)

Arriflex 535 and 535B

Format: 35mm
Magazine sizes: 400′ and 1000′ coaxial
400′ displacement steadicam magazine

For camera and magazine illustrations and threading diagrams, see Figures 6.17–6.20.

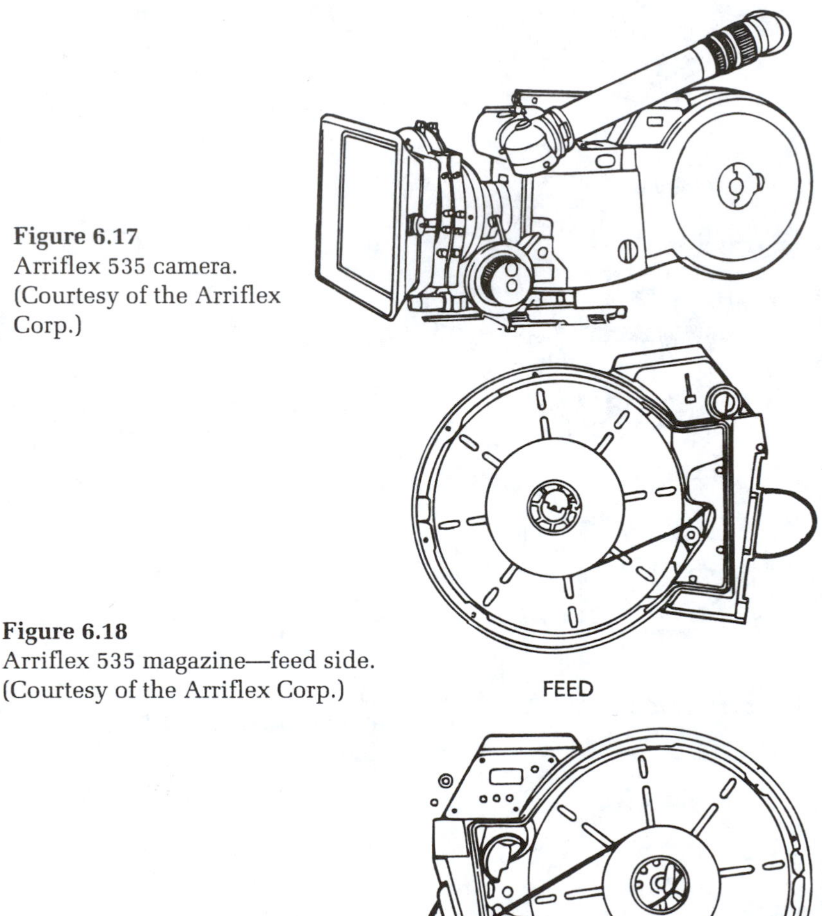

Figure 6.17
Arriflex 535 camera. (Courtesy of the Arriflex Corp.)

Figure 6.18
Arriflex 535 magazine—feed side. (Courtesy of the Arriflex Corp.)

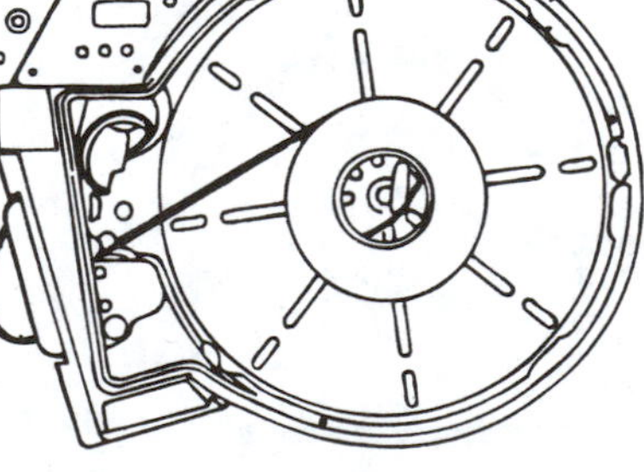

Figure 6.19
Arriflex 535 magazine—take-up side. (Courtesy of the Arriflex Corp.)

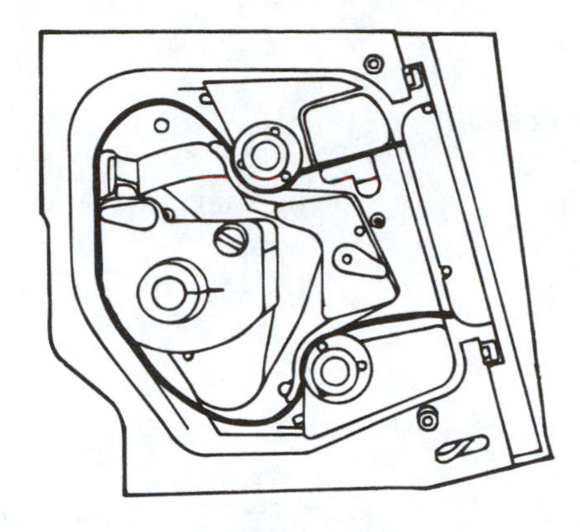

Figure 6.20
Arriflex 535 camera threading. (Courtesy of the Arriflex Corp.)

Arriflex 435

Format: 35mm
Magazine sizes: 400′ and 1000′ displacement

For camera illustration, see Figure 6.21.

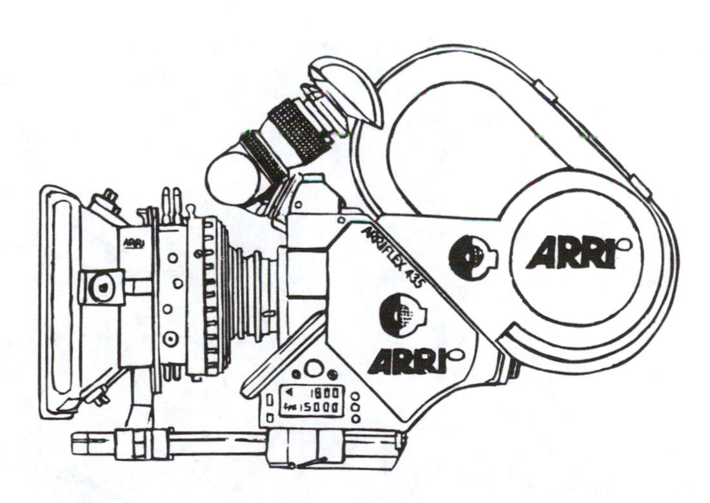

Figure 6.21 Arriflex 435 camera. (Courtesy of the Arriflex Corp.)

Arriflex 35BL3 and 35BL4

Format: 35mm

Magazine sizes: 400′ and 1000′ coaxial

For camera and magazine illustrations and threading diagrams, see Figures 6.22–6.25.

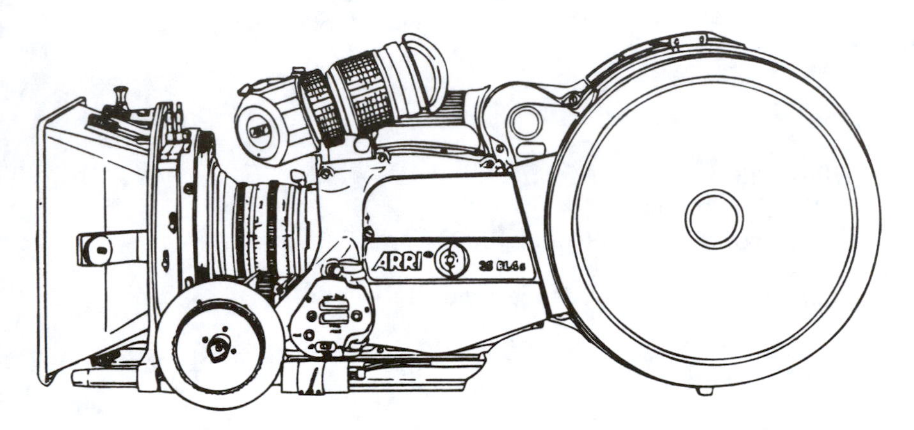

Figure 6.22 Arriflex 35BL camera. (Courtesy of the Arriflex Corp.)

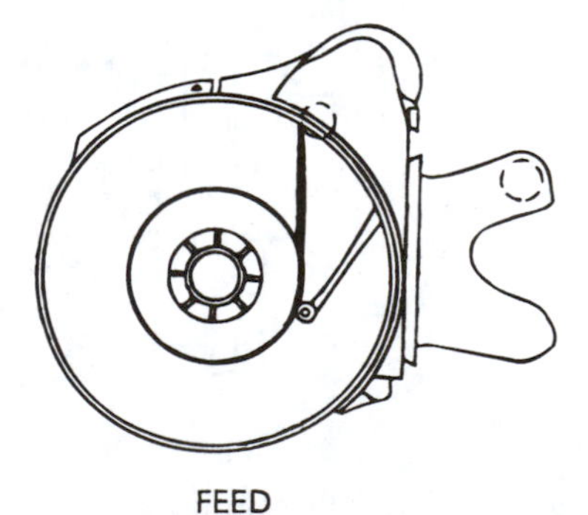

Figure 6.23
Arriflex 35BL magazine—feed side. (Courtesy of the Arriflex Corp.)

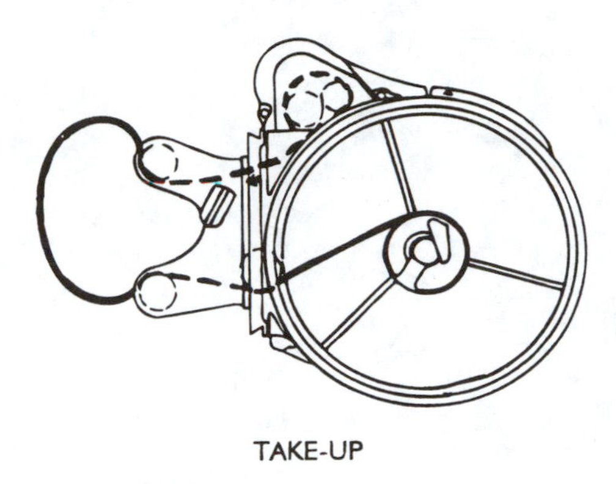

Figure 6.24
Arriflex 35BL magazine—take-up side. (Courtesy of the Arriflex Corp.)

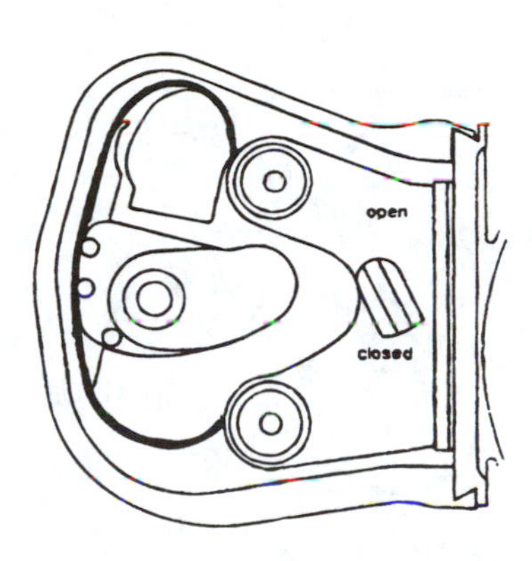

Figure 6.25
Arriflex 35BL camera threading. (Courtesy of the Arriflex Corp.)

Arriflex 35-3

Format: 35mm
Magazine sizes: 200′, 400′ 1000′ displacement
400′ coaxial handheld shoulder magazine
400′ displacement steadicam magazine

For camera and magazine illustrations and threading diagrams, see Figures 6.26–6.32.

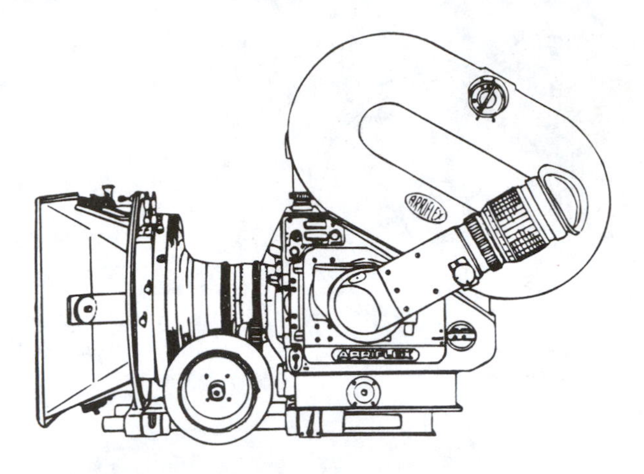

Figure 6.26 Arriflex 35-3 camera. (Courtesy of the Arriflex Corp.)

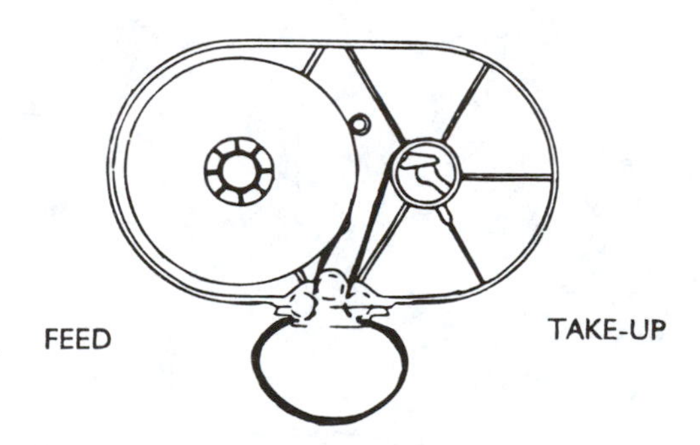

Figure 6.27
Arriflex 35-3, 400′ magazine. (Courtesy of the Arriflex Corp.)

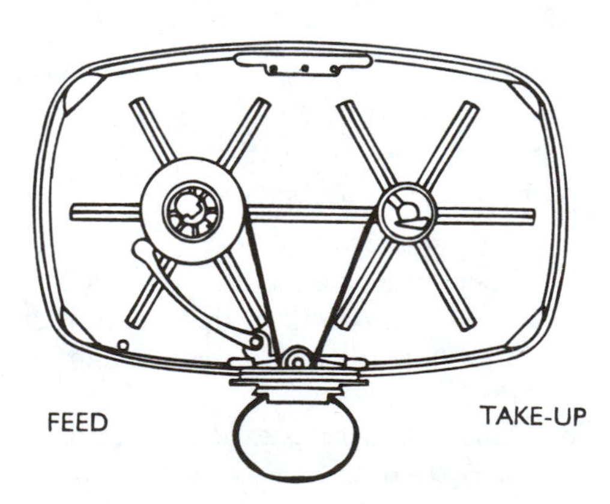

Figure 6.28
Arriflex 35-3, 1000′ magazine. (Courtesy of the Arriflex Corp.)

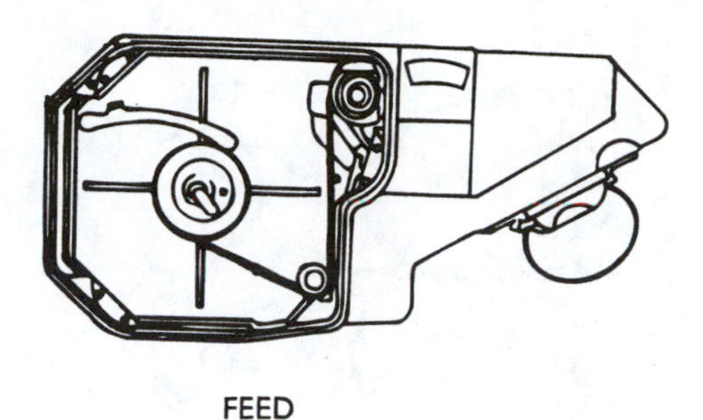

Figure 6.29
Arriflex 35-3 shoulder magazine—feed side. (Courtesy of the Arriflex Corp.)

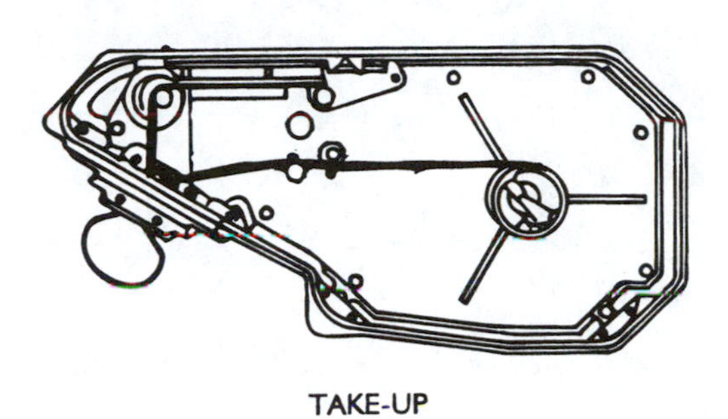

Figure 6.30
Arriflex 35-3 Shoulder Magazine—take-up side. (Courtesy of the Arriflex Corp.)

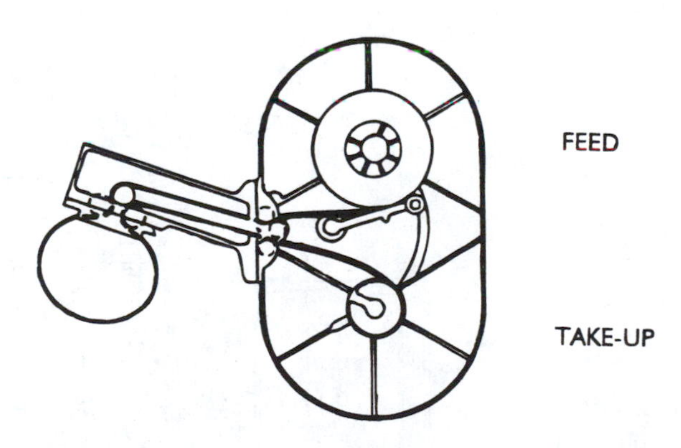

Figure 6.31 Arriflex 35-3 steadicam magazine. (Courtesy of the Arriflex Corp.)

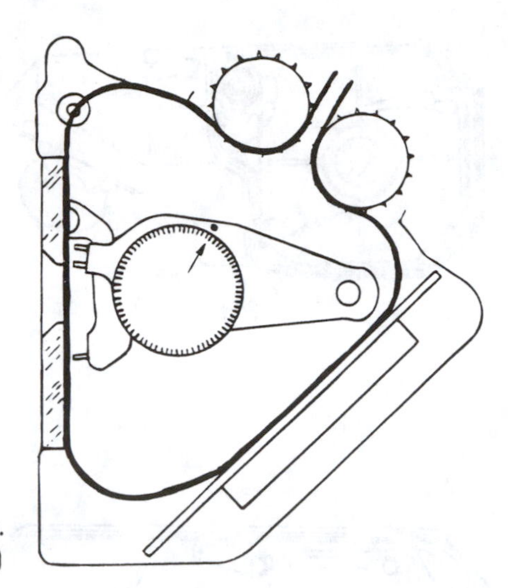

Figure 6.32
Arriflex 35-3 camera threading.
(Courtesy of the Arriflex Corp.)

Arriflex 765

Format: 65mm
Magazine sizes: 500′ and 1000′ displacement

For camera and magazine illustrations and threading diagrams, see Figures 6.33– 6.35.

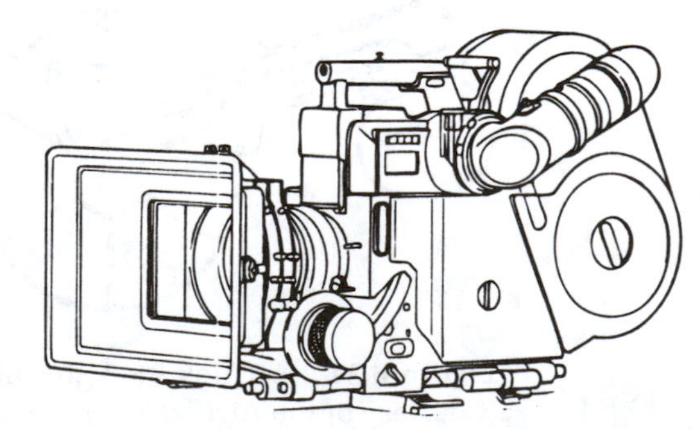

Figure 6.33
Arriflex 765 camera.
(Courtesy of the Arriflex Corp.)

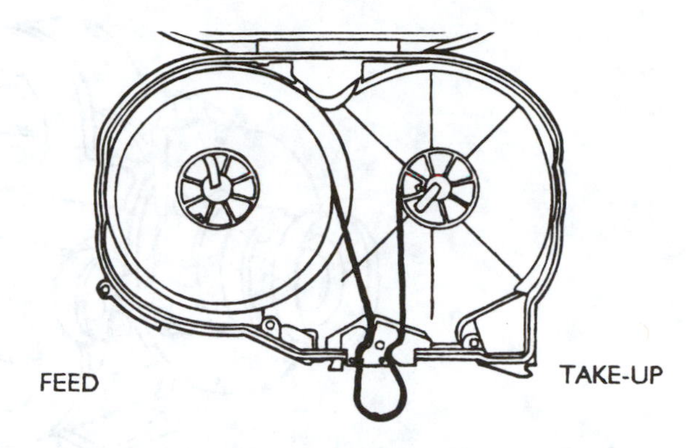

Figure 6.34 Arriflex 765 magazine. (Courtesy of the Arriflex Corp.)

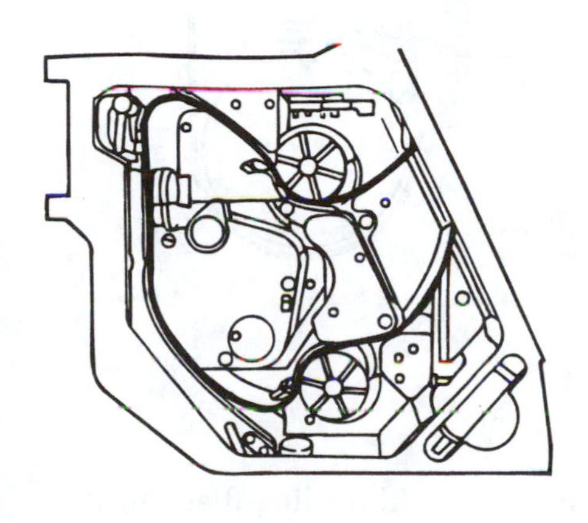

Figure 6.35
Arriflex 765 camera threading. (Courtesy of the Arriflex Corp.)

Bell & Howell Eyemo

Format: 35mm

Magazine sizes: 100′ daylight spool, internal load only

For camera illustration and threading diagrams, see Figures 6.36 and 6.37.

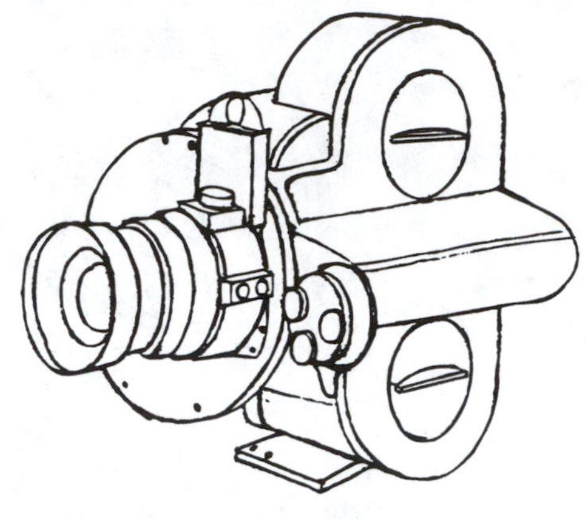

Figure 6.36
Bell and Howell Eyemo camera.

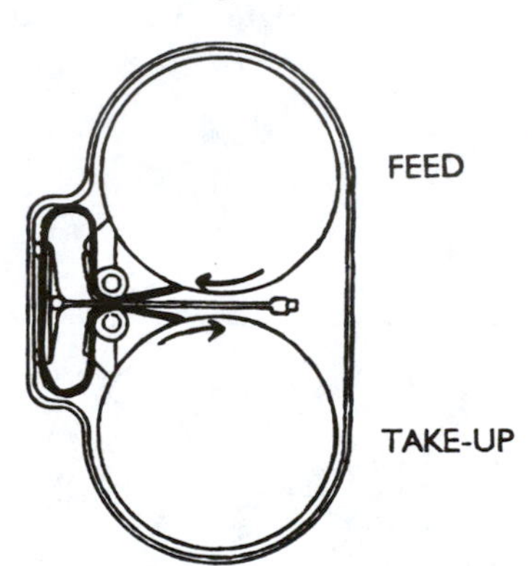

Figure 6.37
Bell and Howell Eyemo threading diagram.

Cinema Products CP16

Format: 16mm
Magazine sizes: 400′ displacement

For camera and magazine illustrations and threading diagrams, see Figures 6.38–6.40.

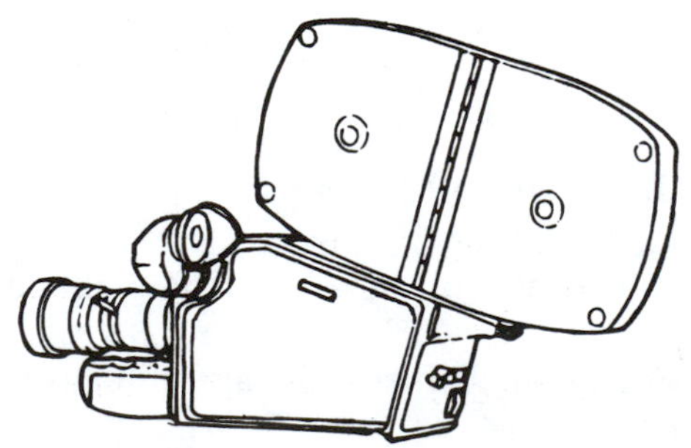

Figure 6.38
Cinema Products CP16 camera.

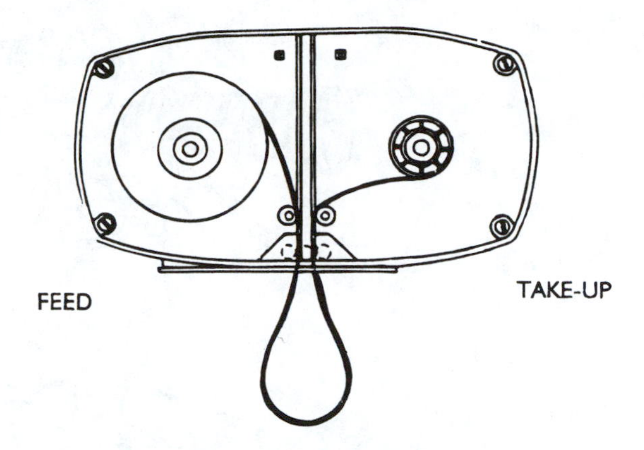

Figure 6.39 Cinema Products CP16 magazine.

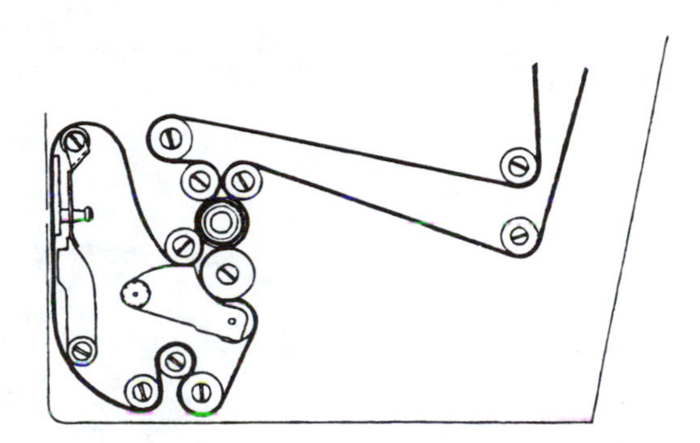

Figure 6.40 Cinema Products CP16 camera threading.

Eclair ACL

Format: 16mm
Magazine sizes: 200′ and 400′ coaxial

For camera and magazine illustrations and threading diagrams, see Figures 6.41–6.42.

Figure 6.41
Eclair ACL camera.

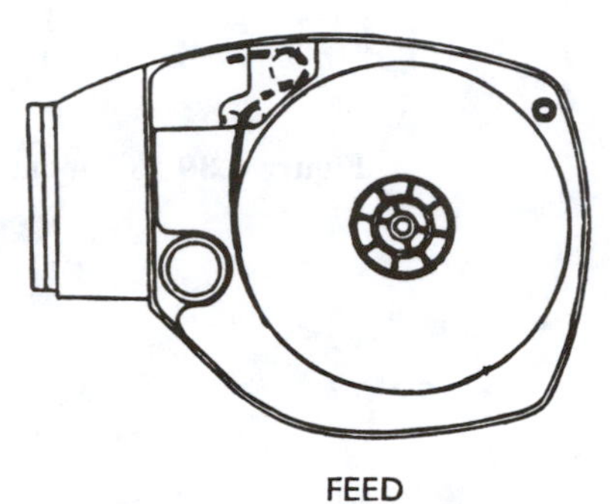

Figure 6.42
Eclair ACL, 400′ magazine—feed side.

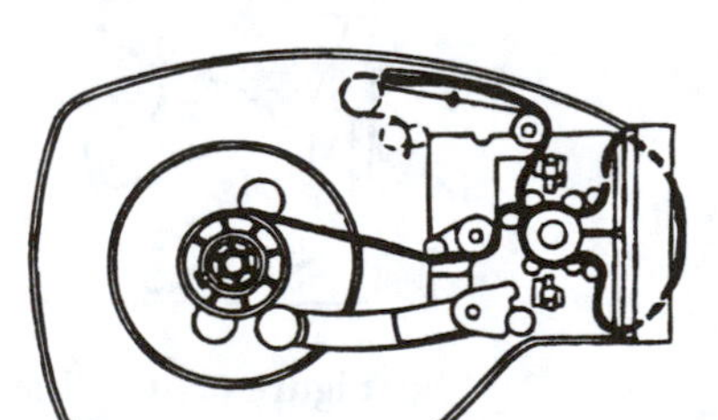

Figure 6.43
Eclair ACL, 400′ magazine—take-up side.

Eclair NPR

Format: 16mm
Magazine sizes: 400′ coaxial

For camera and magazine illustrations and threading diagrams, see Figures 6.44–6.46.

Figure 6.44
Eclair NPR camera.

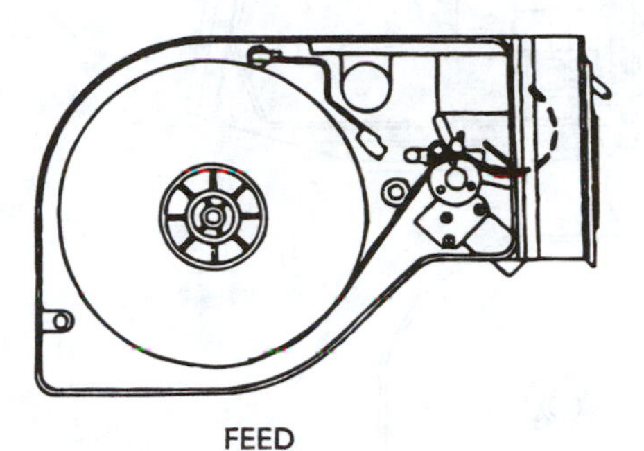

Figure 6.45
Eclair NPR magazine—feed side.

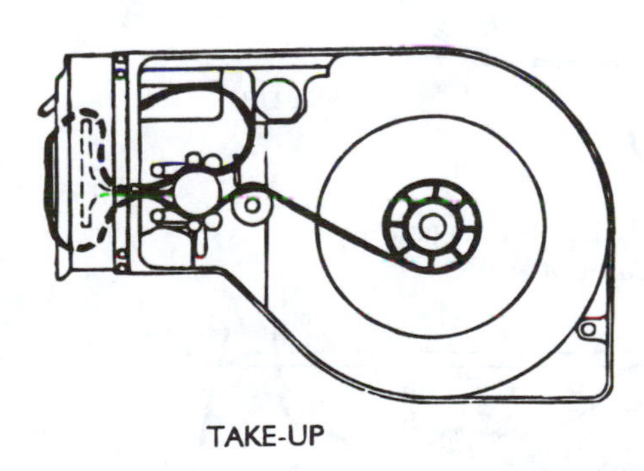

Figure 6.46
Eclair NPR magazine—take-up side.

Leonetti Ultracam

Format: 35mm
Magazine sizes: 500′ and 1000′ displacement

For camera and magazine illustrations and threading diagrams, see Figures 6.47–6.49.

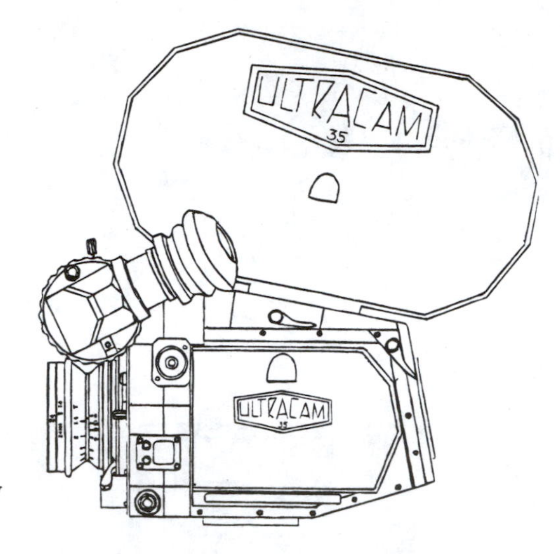

Figure 6.47 Leonetti Ultracam. (Courtesy of Leonetti Camera.)

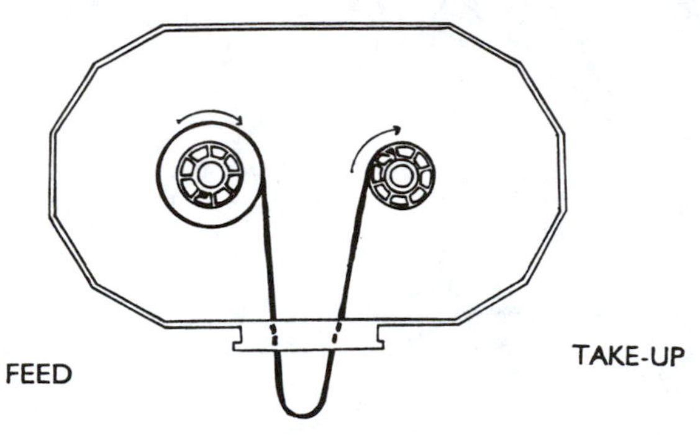

Figure 6.48 Leonetti Ultracam magazine. (Courtesy of Leonetti Camera.)

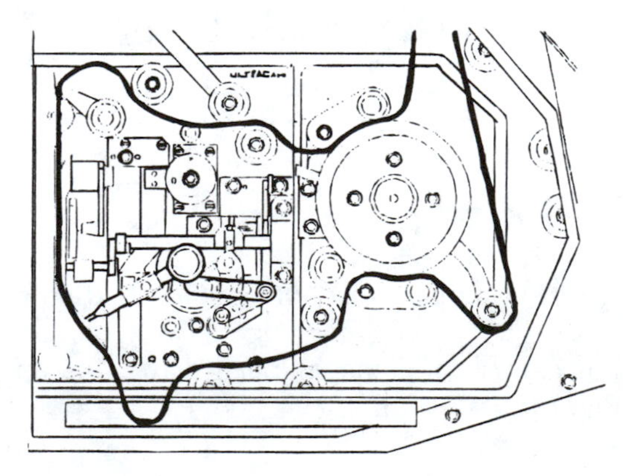

Figure 6.49 Leonetti Ultracam camera threading. (Courtesy of Leonetti Camera.)

Moviecam Compact and Moviecam Super America

Format: 35mm

Magazine sizes: 500′ and 1000′ displacement

For camera and magazine illustrations and threading diagrams, see Figures 6.50–6.55.

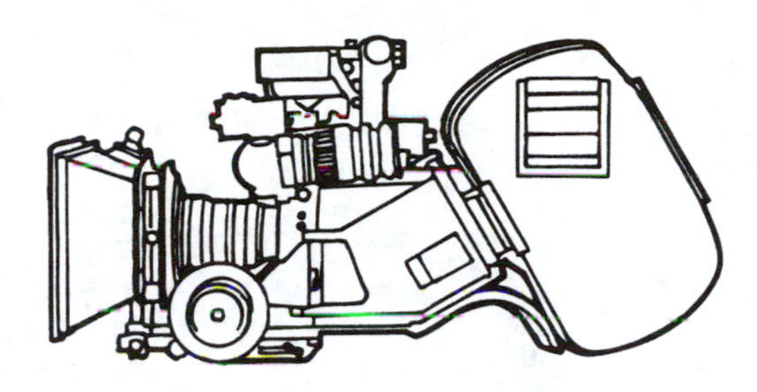

Figure 6.50
Moviecam compact camera.

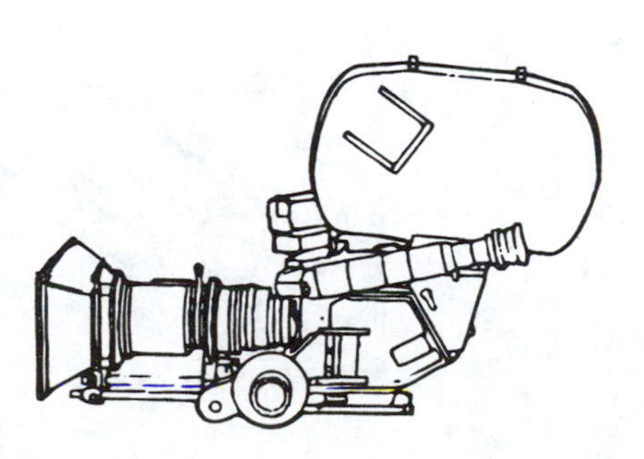

Figure 6.51
Moviecam Super America camera.

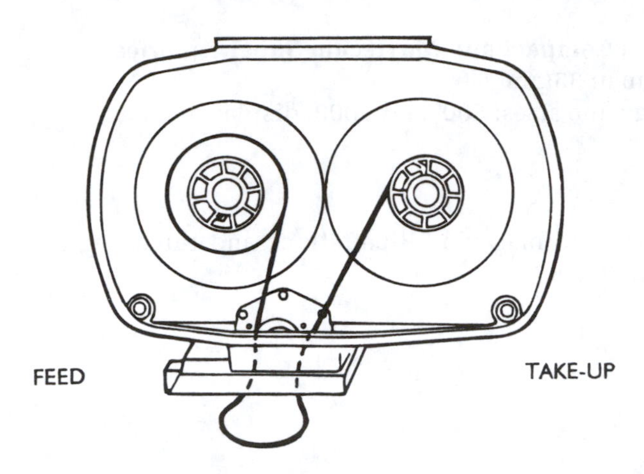

Figure 6.52 Moviecam magazine.

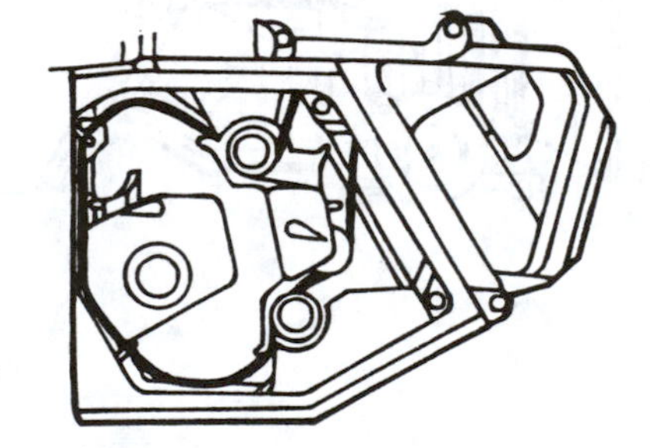

Figure 6.53
Moviecam compact camera threading—top load.

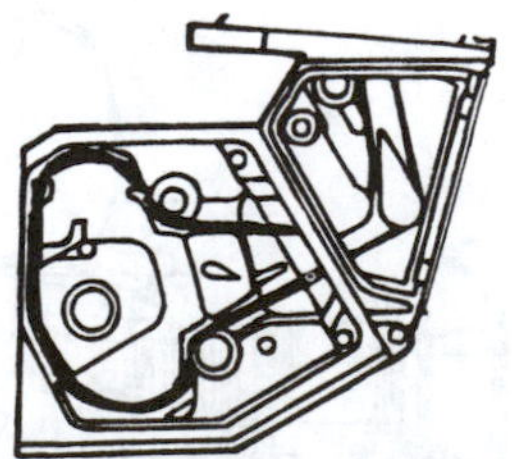

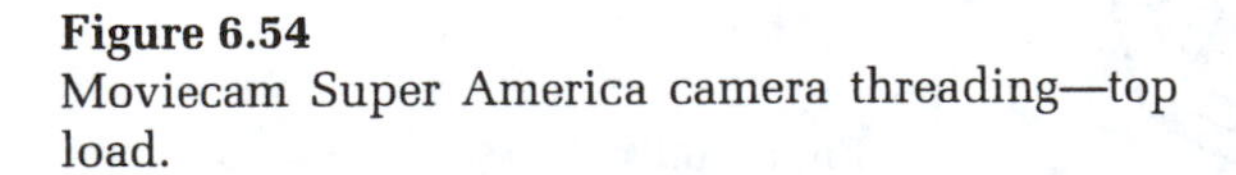

Figure 6.54
Moviecam Super America camera threading—top load.

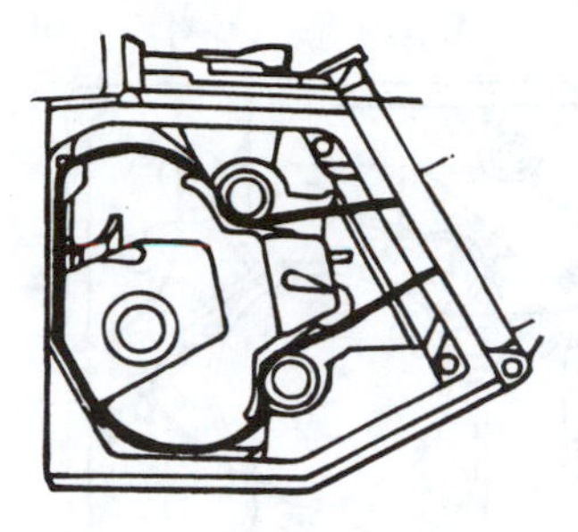

Figure 6.55
Moviecam camera threading—slant load.

Panavision Panaflex-16

Format: 16mm

Magazine sizes: 500′ and 1200′ displacement

For camera and magazine illustrations and threading diagrams, see Figures 6.56–6.58.

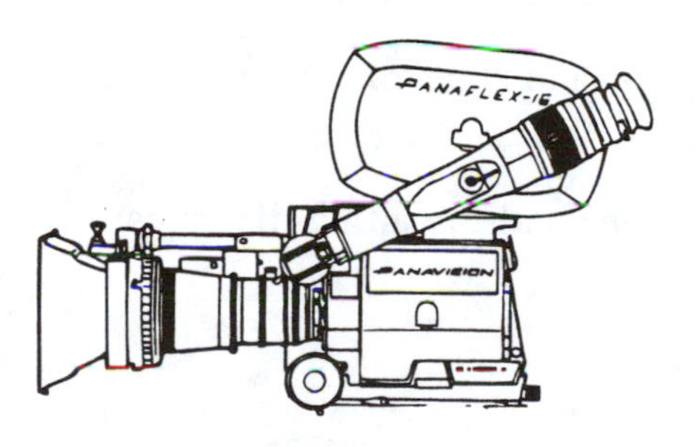

Figure 6.56
Panavision Panaflex-16 camera. (Courtesy of Panavision, Inc.)

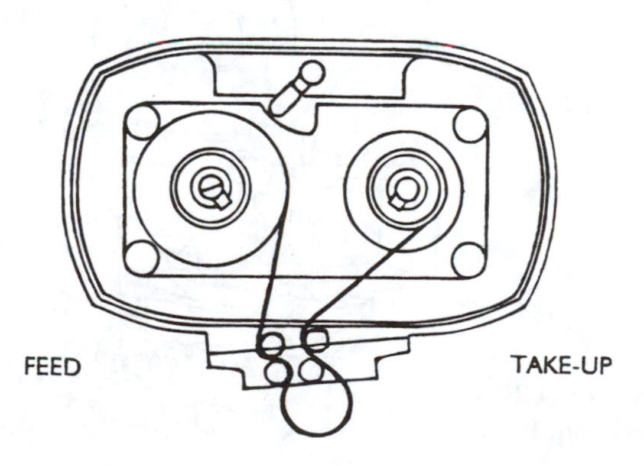

Figure 6.57
Panavision Panaflex-16 magazine. (Courtesy of Panavision, Inc.)

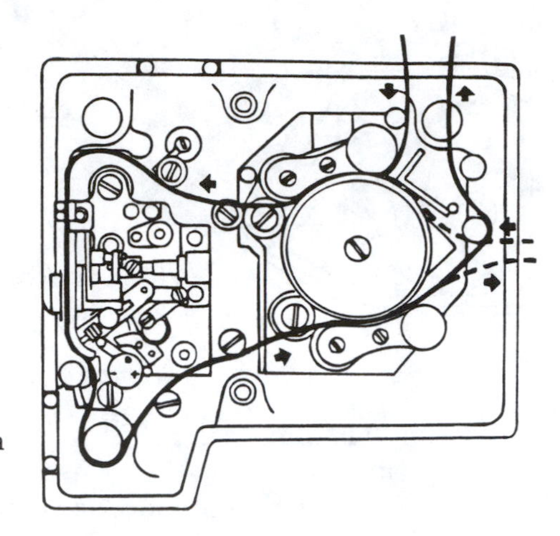

Figure 6.58
Panavision Panaflex-16 camera threading. (Courtesy of Panavision, Inc.)

Panavision Panaflex Golden and GII

Format: 35mm
Magazine sizes: 250′, 500′, 1000′, and 2000′ displacement

For camera and magazine illustrations and threading diagrams, see Figures 6.59–6.61

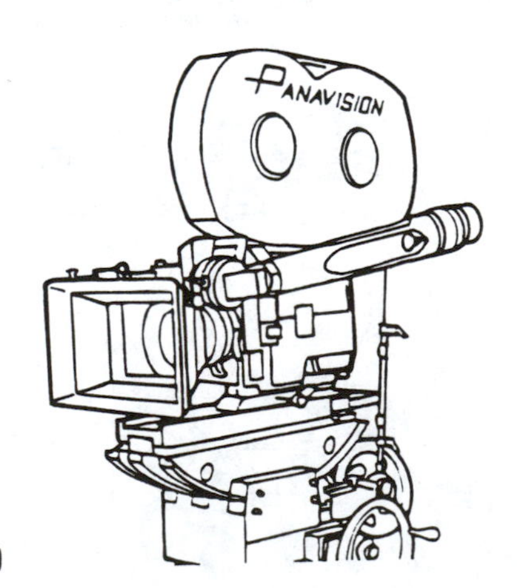

Figure 6.59
Panavision Panaflex Golden and GII camera. (Courtesy of Panavision, Inc.)

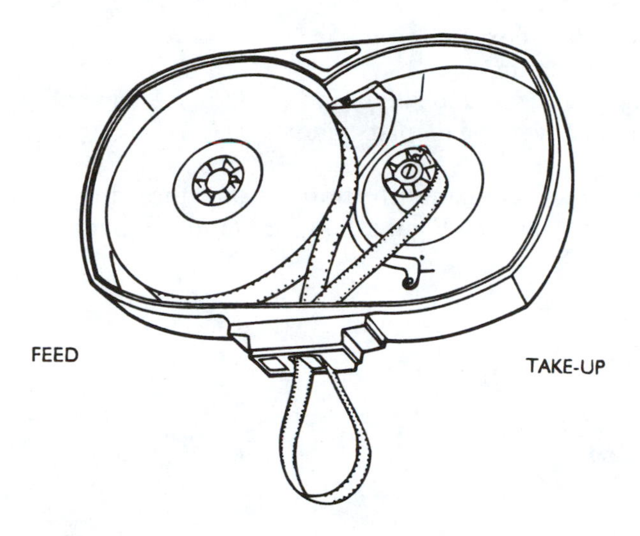

Figure 6.60 Panavision Panaflex 35mm magazine—Golden, GII, Platinum, Panaflex-X, and Panastar. (Courtesy of Panavision, Inc.)

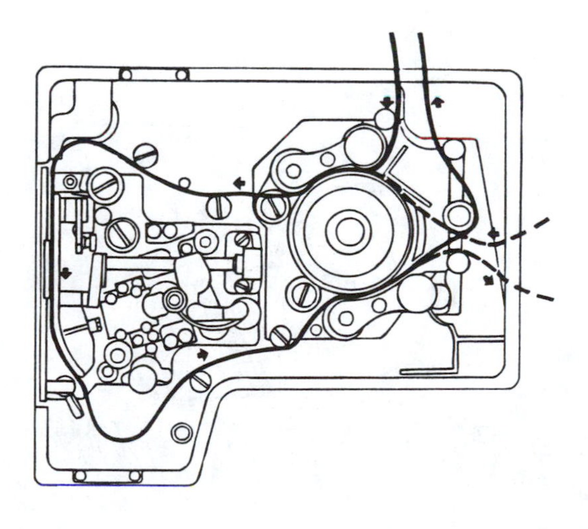

Figure 6.61
Panavision Panaflex Golden and GII camera threading. (Courtesy of Panavision, Inc.)

Panavision Panaflex Platinum

Format: 35mm

Magazine sizes: 250′, 500′, and 1000′ displacement
1000′ reversing displacement

For camera and magazine illustrations and threading diagrams, see Figures 6.62–6.65. See Figure 6.60 for standard magazine threading diagram.

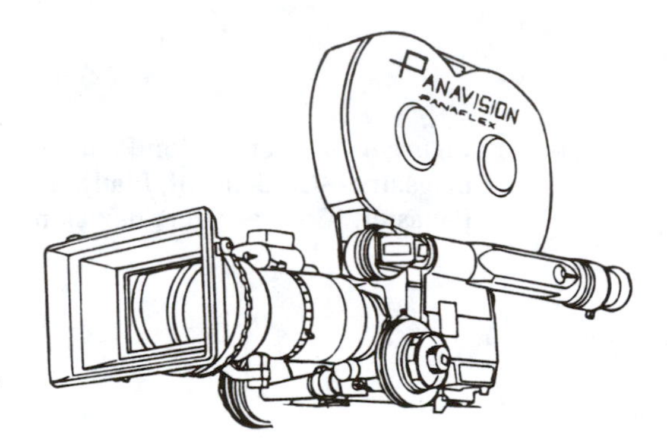

Figure 6.62
Panavision Panaflex Platinum camera. (Courtesy of Panavision, Inc.)

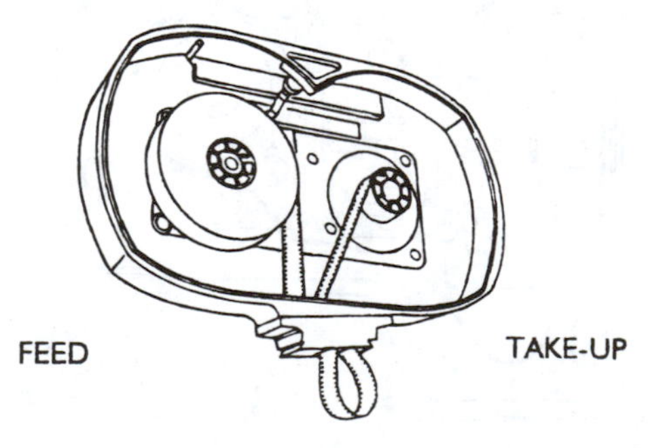

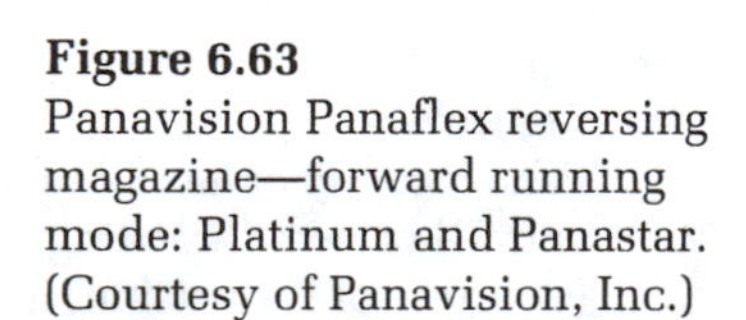

Figure 6.63
Panavision Panaflex reversing magazine—forward running mode: Platinum and Panastar. (Courtesy of Panavision, Inc.)

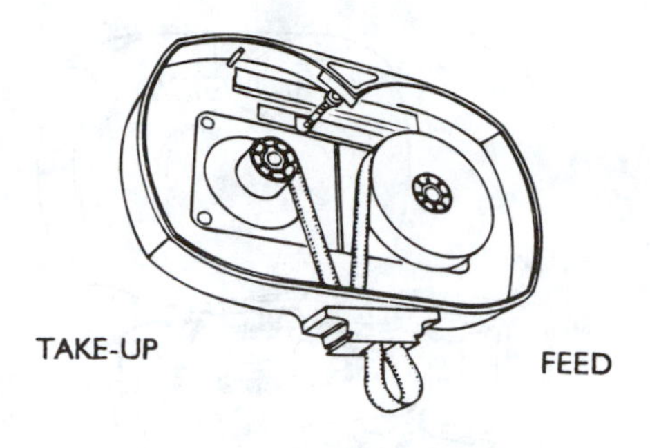

Figure 6.64
Panavision Panaflex reversing magazine—reverse running mode: Platinum and Panastar. (Courtesy of Panavision, Inc.)

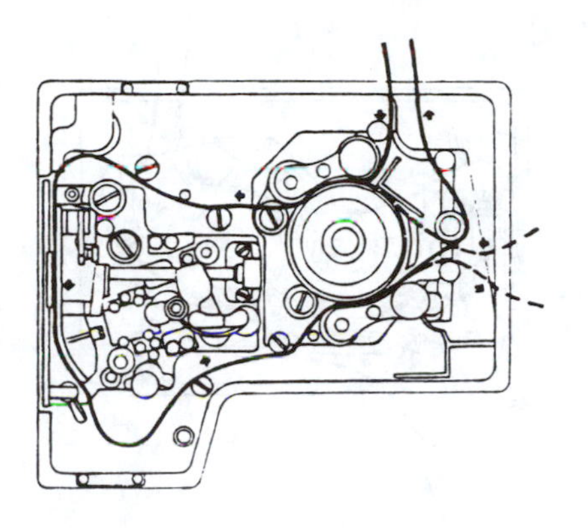

Figure 6.65
Panavision Panaflex Platinum camera threading. (Courtesy of Panavision, Inc.)

Panavision Panaflex-X

Format: 35mm

Magazine sizes: 250′, 500′, 1000′, and 2000′ displacement

For camera and magazine illustrations and threading diagrams, see Figures 6.66 and 6.67 See Figure 6.60 for standard magazine threading diagram.

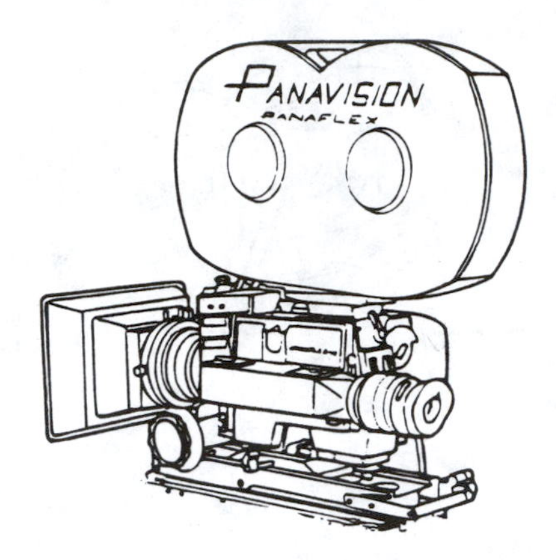

Figure 6.66
Panavision Panaflex-X camera. (Courtesy of Panavision, Inc.)

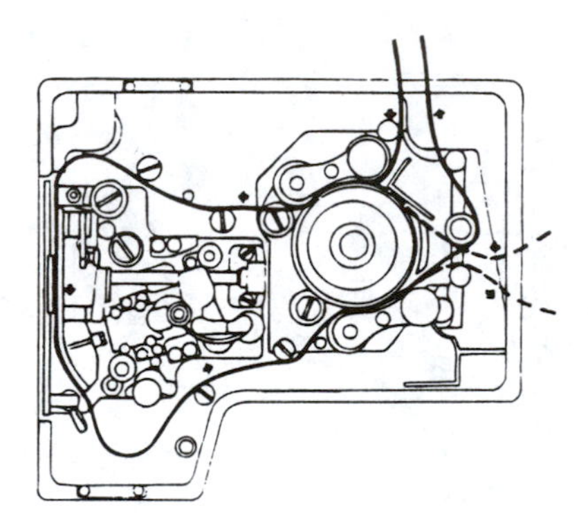

Figure 6.67
Panavision Panaflex-X camera threading. (Courtesy of Panavision, Inc.)

Panavision Panastar I and Panastar II

Format: 35mm high speed
Magazine sizes: 500′ and 1000′ displacement
Panastar II only: 1000′ reversing displacement

For camera and magazine illustrations and threading diagrams, see Figures 6.68 and 6.69. See Figure 6.60 for standard magazine threading diagram. See Figures 6.63 and 6.64 for reversing magazine threading diagrams.

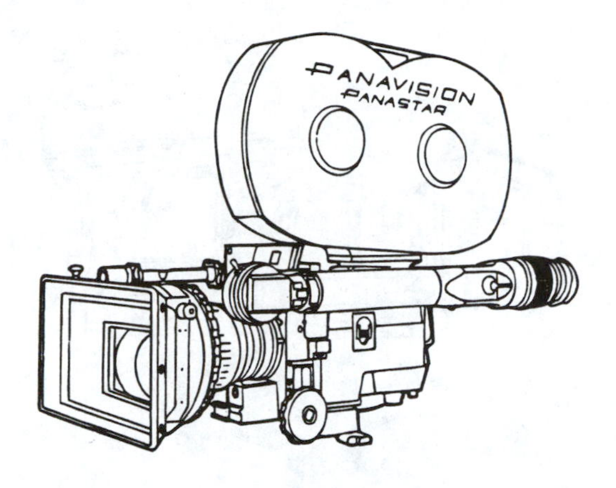

Figure 6.68
Panavision Panastar camera. (Courtesy of Panavision, Inc.)

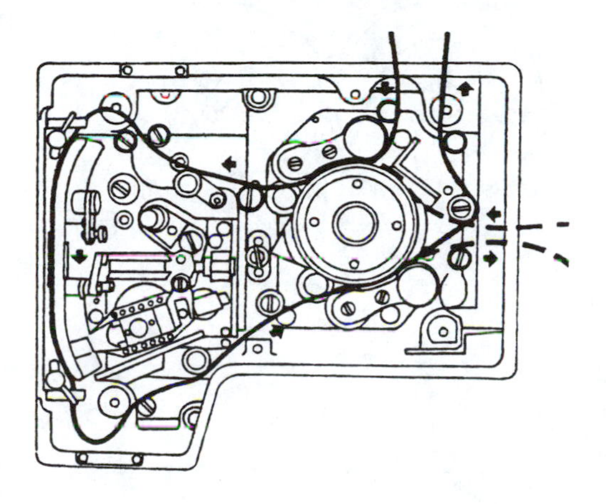

Figure 6.69
Panavision Panastar camera threading. (Courtesy of Panavision, Inc.)

Panavision Super PSR

Format: 35mm
Magazine sizes: 1000′ double chamber displacement.

For camera and magazine illustrations and threading diagrams, see Figures 6.70–6.72.

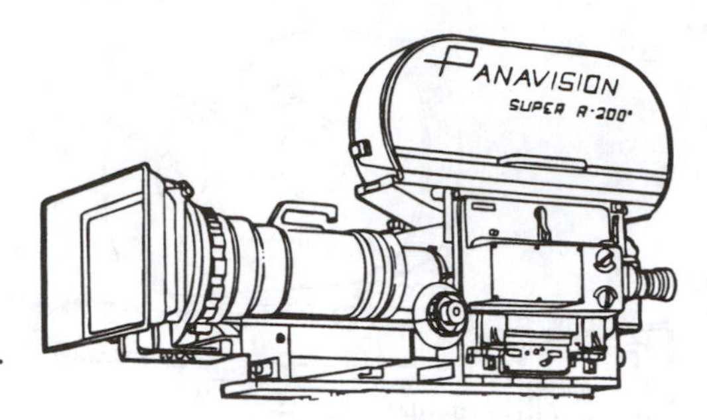

Figure 6.70
Panavision Super PSR camera. (Courtesy of Panavision, Inc.)

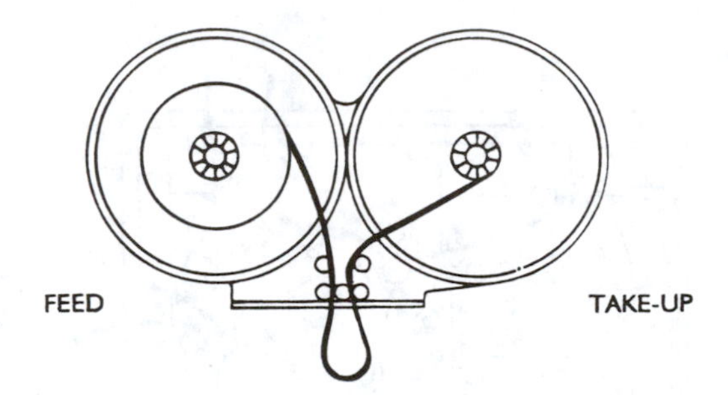

Figure 6.71
Panavision Super PSR magazine. (Courtesy of Panavision, Inc.)

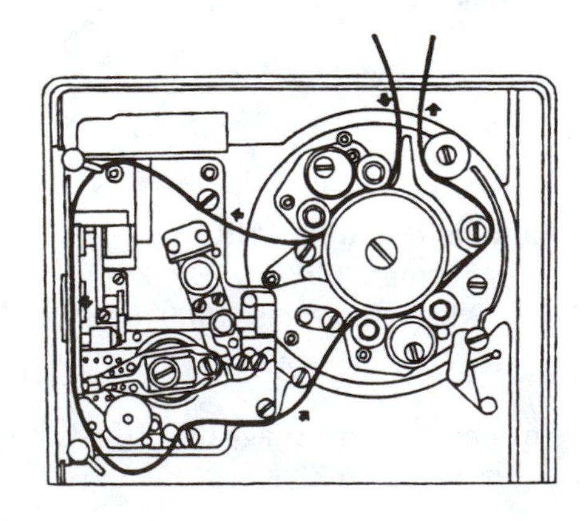

Figure 6.72
Panavision Super PSR camera threading. (Courtesy of Panavision, Inc.)

Panavision System 65

Format: 65mm

Magazine sizes: 500′ and 1000′ displacement

For camera and magazine illustrations and threading diagrams, see Figures 6.73–6.75.

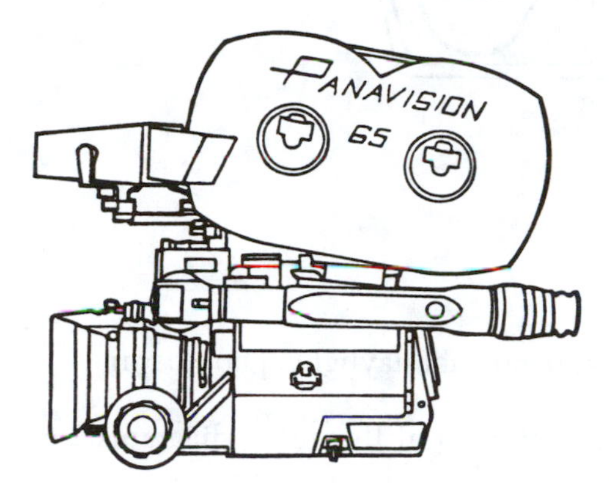

Figure 6.73
Panavision System 65 camera. (Courtesy of Panavision, Inc.)

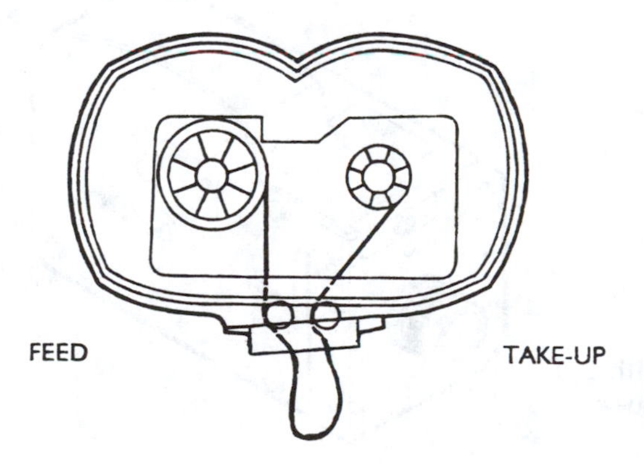

Figure 6.74
Panavision System 65 magazine. (Courtesy of Panavision, Inc.)

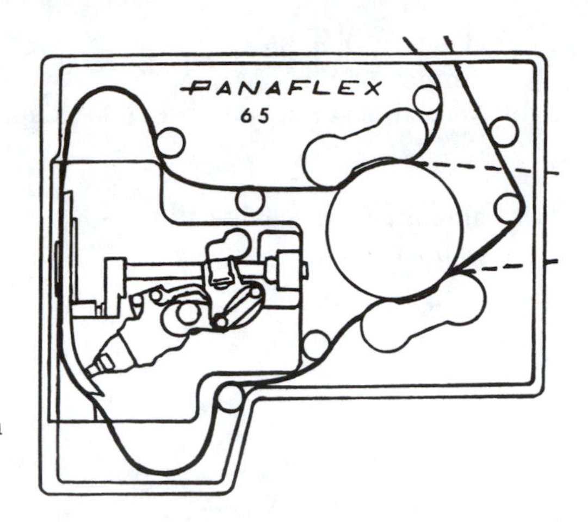

Figure 6.75
Panavision System 65 camera threading. (Courtesy of Panavision, Inc.)

Photo-Sonics 1PL

Format: 16mm high speed
Magazine sizes: 200′ and 400′ coaxial daylight spool loading

For camera and magazine illustrations and threading diagrams, see Figures 6.76 and 6.77.

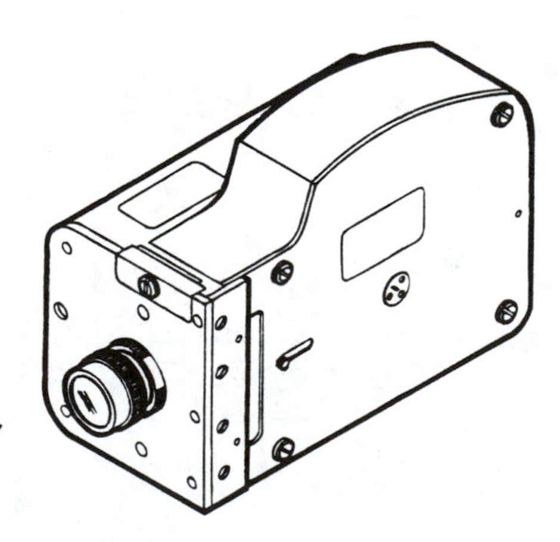

Figure 6.76
Photo-Sonics 1PL camera with 400′ magazine. (Courtesy of Photo-Sonics, Inc.)

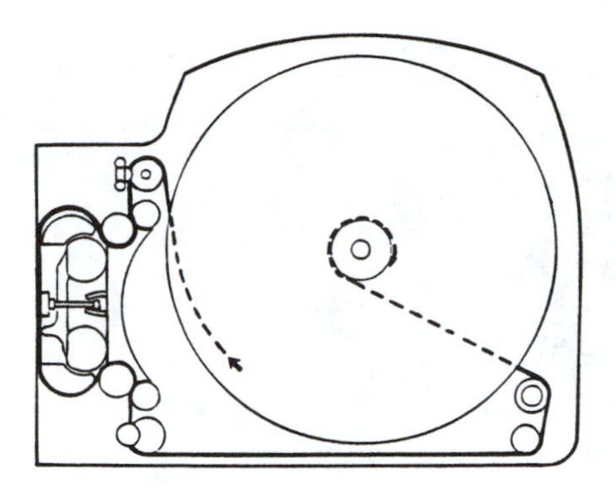

Figure 6.77
Photo-Sonics 1PL magazine—film path outline. (Courtesy of Photo-Sonics, Inc.)

Photo-Sonics 35-4B/4C

Format: 35mm high speed
Magazine sizes: 1000′ displacement

For camera and magazine illustrations and threading diagrams, see Figures 6.78–6.81.

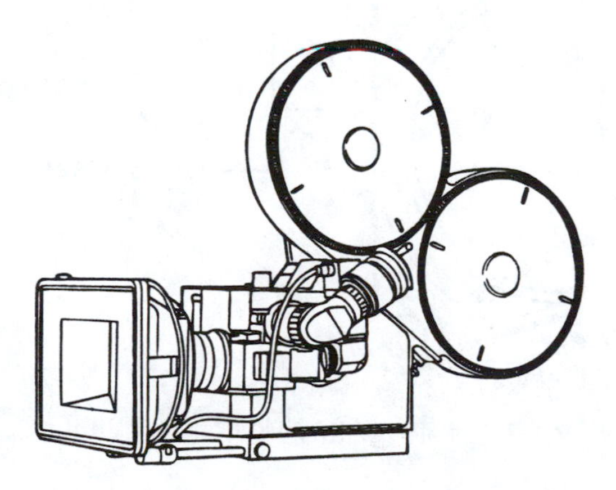

Figure 6.78
Photo-Sonics 35-4B/4C camera. (Courtesy of Photo-Sonics, Inc.)

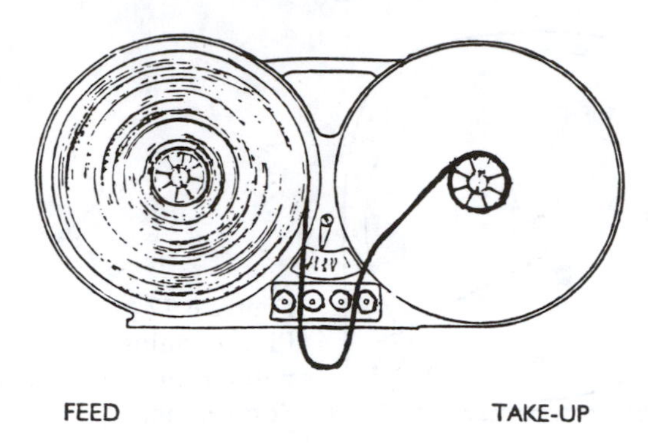

Figure 6.79 Photo-Sonics 35-4B/4C magazine. (Courtesy of Photo-Sonics, Inc.)

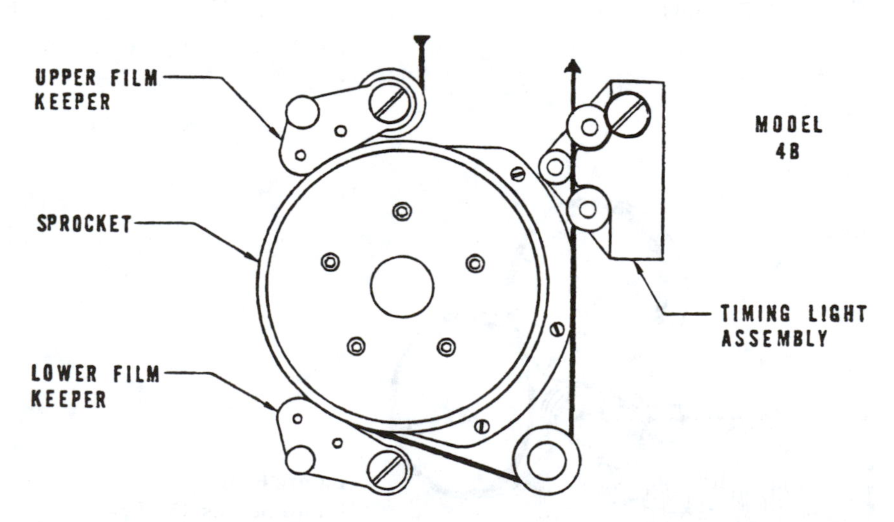

Figure 6.80 Photo-Sonics 35-4B camera threading. (Courtesy of Photo-Sonics, Inc.)

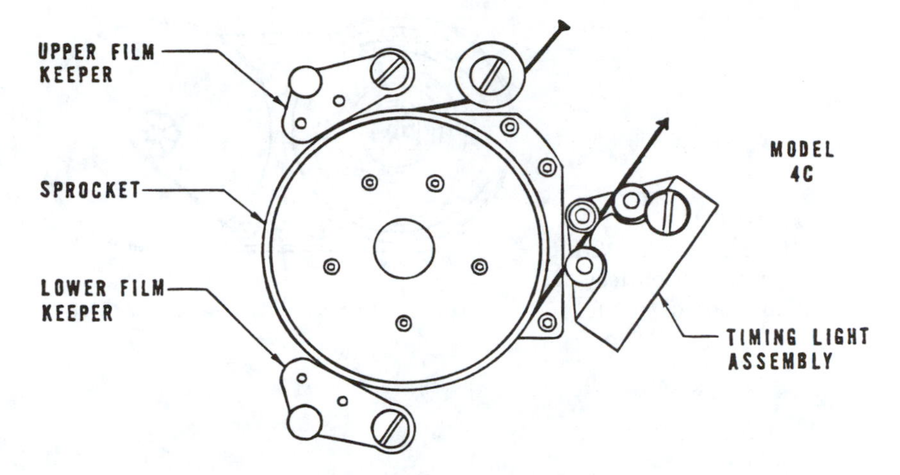

Figure 6.81 Photo-Sonics 35-4C camera threading. (Courtesy of Photo-Sonics, Inc.)

Photo-Sonics 35-4E/ER

Format: 35mm high speed

Magazine sizes: 1000′ displacement

For camera and magazine illustrations and threading diagrams, see Figures 6.82–6.84.

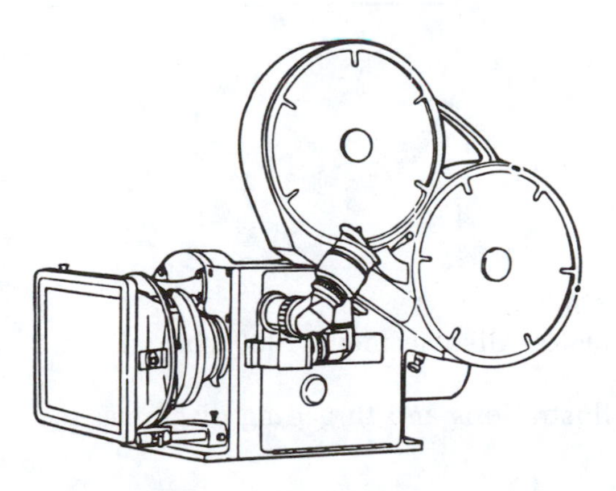

Figure 6.82 Photo-Sonics 35-4E/ER camera. (Courtesy of Photo-Sonics, Inc.)

Figure 6.83
Photo-Sonics 35-4E/ER magazine. (Courtesy of Photo-Sonics, Inc.)

FEED

TAKE-UP

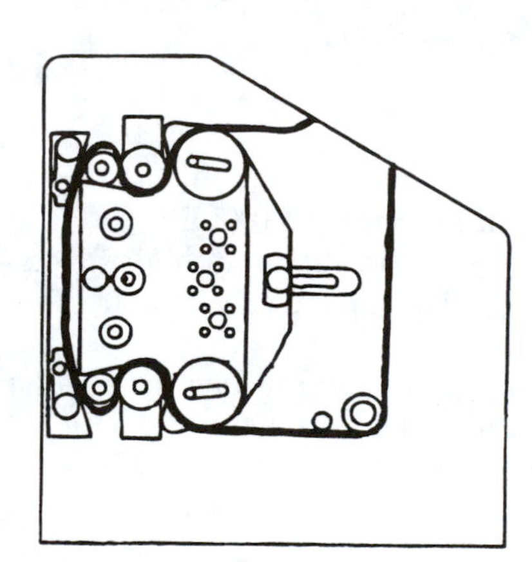

Figure 6.84
Photo-Sonics 35-4E/ER camera threading. (Courtesy of Photo-Sonics, Inc.)

Photo-Sonics 35-4ML

Format: 35mm high speed

Magazine sizes: 200′ and 400′ displacement; 1000′ coaxial

For camera and magazine illustrations and threading diagrams, see Figures 6.85 and 6.86.

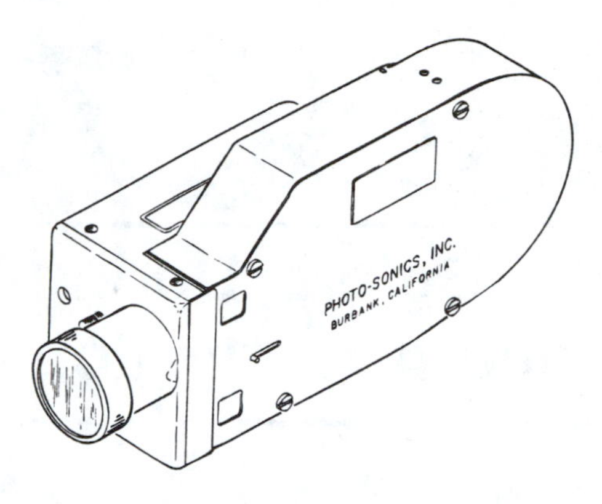

Figure 6.85
Photo-Sonics 35-4ML camera with 400′ magazine. (Courtesy of Photo-Sonics, Inc.)

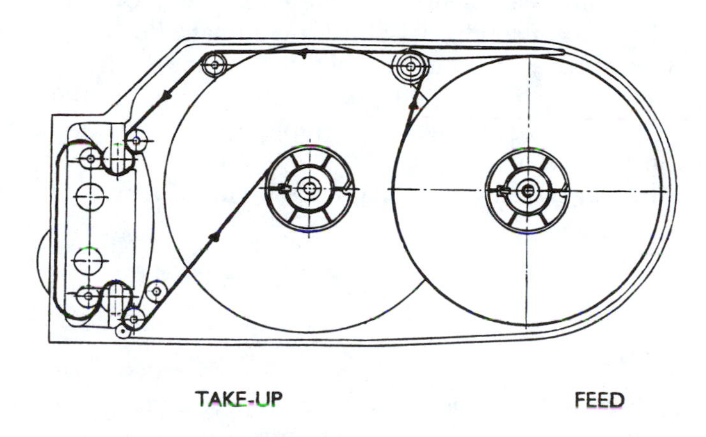

Figure 6.86 Photo-Sonics 35-4ML 400′ magazine. (Courtesy of Photo-Sonics, Inc.)

7

Before, During and After the Job

Now that you have read the first six chapters, and know how to do the job of a First and Second Assistant Cameraman, I'd like to mention some of the things you should do before you get the job, how to act once you have the job, and finally what to do when it's all over. Some of what is discussed here includes preparing a resume, questions to ask during the interview, proper set etiquette, and how to behave while on the job, and staying in contact with crew people after the job.

BEFORE THE JOB

The Resume

One of the first things you should do is prepare a resume. At the very beginning you will most likely have minimal experience. If you have recently graduated from film school, you will probably have some experience on student productions. A beginning resume should list any production experience that you may have. This includes production assistant, craft service, grip, electrician, and any other jobs you may have done. At the top of the resume you should state that your goal is to work in the camera department so that anyone reading it will know that you do have a specific goal. Once you have acquired more camera-related experience, then you may remove the other jobs not related to the camera department.

Your resume should include your basic personal contact information: name, address and telephone numbers. If you have a pager or fax machine, be sure to include these numbers also. I recommend getting a pager as soon as possible so you don't miss out on any job calls. As a freelance camera assistant, prospective employers need to be able to get in touch with you.

Next, your resume should list your production credits. These are most often listed in reverse chronological order, which means that the most recent job is at the top of the listing. The exception to this is if you have any production credits from well-known, recognizable productions. In this case, those credits should be listed first. As a Director of Photography (D.P.) or Production Manager looks at your resume, these names will jump out at them and indicate that you are qualified for the job.

The format that you use for the resume is up to your personal preference. Most resumes that I have seen contain the same or similar basic information. This includes the title of the production, type of production—feature film, television show, commercial, etc.—whether the job performed was that of a 1st A.C., 2nd A.C., or Loader, the name of the D.P., and sometimes the name of the Director, Producer, or production company.

Many production managers or D.P.s who you interview with will most often ask, "What D.P.s have you worked with?" In preparing my resume, I included the names of all D.P.s that I have worked with on the various productions. I currently work as both a Camera Operator and First Assistant, so I list all of my production credits for these positions as well as my past experience as a Second Assistant. I have listed my credits in sub-categories based on the type of production—television series, feature film, commercial, music video, and other credits.

Following your listing of credits, you may list any special skills or equipment knowledge that relates to your experience. Following this, you should list your education, including the name of the school, years attended, and degree earned. At the end of the resume, the following statement should be included: "References available upon request." Don't volunteer reference information unless it is asked for. When giving names of references, be sure that you have permission from the person whose name you are giving out, beforehand.

The most important thing to remember about your resume is don't lie. If you do, it will be discovered sooner or later and will only cause you more problems than it is worth. An abbreviated version of my resume is shown in Figure 7.1 to give you an idea of how to set up your resume.

Your resume is done, and now you need to get that first job. Send it out to as many production companies as possible. The best places to

David E. Elkins
123 Main Street
Anytown, USA

Home (123) 555-1234
Fax (123) 555-6789
Pager (800) 123-4567

FIRST ASSISTANT CAMERAMAN

EXPERIENCE

TELEVISION

The Wonder Years	D.P. - Steve Confer
Star Trek - The Next Generation	D.P. - Marvin Rush
Gabriel's Fire	D.P. - Victor Goss
Pro's & Cons	D.P. - Victor Goss
Dark Justice	D.P. - Tony Palmieri
P.S. I Luv U (B Camera)	D.P. - Bob Hayes

FEATURE FILMS

Claire of the Moon	D.P. - Randy Sellars
Prey of the Chameleon	D.P. - Randy Sellars
Red Surf	D.P. - John Schwartzman
The Unholy (2nd Unit)	D.P. - John Schwartzman
Waxwork	D.P. - Gerry Lively

COMMERCIALS

Life Savers	D.P. - Aaron Schneider
McDonalds	D.P. - Victor Goss
Carl's Junior	D.P. - Randy Sellars
Minute Maid Orange Juice (B Camera)	D.P. - Rolf Kesterman
CBN - The Family Channel	D.P. - Marvin Rush

MUSIC VIDEOS

Red Hot Chili Peppers - "Under the Bridge"	D.P. - Eric Edwards
Air Supply - "Stop the Tears"	D.P. - Tim Roarke
Beastie Boys - "Pass the Mic"	D.P. - Tim Roarke
Jackal - "Down on Me"	D.P. - Eagle Egilsson
Coolio - "County Line"	D.P. - Eagle Egilsson

OTHER

Pediatric CPR - Industrial	D.P. - Randy Sellars
It's A Blast - PSA	D.P. - Jose Luis Becerra
Keep Them Out of Reach - PSA	D.P. - Eagle Egilsson

Complete knowledge of most professional 16mm and 35mm camera systems.

EDUCATION

B.A. - Motion Picture Production	Columbia College-Hollywood, Hollywood, California, 1985
B.A. - Mathematics	University of Connecticut, Storrs, Connecticut, 1976

References available upon request

Figure 7.1 Example of a resume for a camera assistant.

look for listings are the film industry related trade papers or magazines, such as *Drama Logue, Daily Variety*, and *The Hollywood Reporter*. These publications are available on most newsstands and also by subscription. Each week they will list current productions, along with productions in the pre-production or planning stages.

When mailing your resume, you should include a brief cover letter that introduces yourself and explains why you are writing to the company. Mail your letter and resume to any of the productions that interest you. If there is a telephone number listed, wait about a week and then call them. Ask if your resume was received, and ask if you can come in for a personal interview. Sometimes you may have to work for little or no money to prove yourself. Don't be discouraged by rejection. Be persistent and eventually you will get that first big break.

The Job Interview

Now that you have prepared your resume and sent it out, you are ready to go on that first job interview. In most cases the interview will be conducted by the Director of Photography or Production Manager, or both. Arrive a little early for the interview and be prepared. Have additional copies of your resume with you in case they are asked for. An important part of the interview is asking the right questions. There are many things that you need to know about the job before starting, and you have the right to ask these questions. The following are some key questions that you should ask when interviewing for a job on any production. They are listed in no specific order.

- What format is the film being shot in—16mm or 35mm?
- What camera system will be used?
- Is it a union or non-union crew?
- What is the daily rate for the position I am applying for?
- Is this a flat rate or is there overtime pay after a specific number of hours?
- Do you pay a box or kit rental?
- Is the shooting local or on a distant location?
- If it is a distant location, do you pay travel expenses, per diem, and lodging?
- Are meals provided?
- How many weeks of shooting will there be?
- Are the work weeks 5 or 6 days? Never work a 7 day week.

- How many hours per day do you anticipate shooting? Twelve is good, anything over 12 is too much.
- What are the start and end dates of the shooting schedule?
- Are there any other crew positions still available? Recommend other crew members that you have worked with in the past.

These are some of the typical questions that you should ask during the interview. You may add others as you gain more experience. Once you have completed this basic part of the interview, you may be asked to sign a deal memo, which outlines the terms and conditions of your employment as well as the pay scale. Be sure to obtain a copy of all paperwork that you sign so that if there are any problems or questions later on you can refer to it.

If you have any special camera-related equipment that may be used on the production, ask if the company will rent it from you instead of from the camera rental company. Many assistants own camera batteries, filters, or other camera equipment. Give the production company a fair price to make it worth renting from you. One thing you should remember when owning and renting your equipment is that you are taking business away from a rental house that you will be dealing with regularly in the course of your career. Don't jeopardize your reputation with a rental company just to make a few extra dollars.

DURING THE JOB

Now that you have the job, you should follow some basic guidelines while on the set. There is a proper set etiquette that should be followed by all crew members on any production. How you conduct yourself is just as important as knowing how to do the job properly. You are a professional and should act accordingly.

When you get that first big break and have been hired for your first film job, don't let it go to your head. In other words, don't develop a big ego. Just because you have the basic knowledge to do the job, doesn't mean that you know everything. I learn something new on every job I do. If you make a mistake, admit it. Never try to blame your errors on someone else. And don't forget to thank other crew members for their efforts and help.

Your first time on the set you may feel like a stranger in a foreign land. Learn other crew members names as quickly as possible. If you have any questions, don't hesitate to ask. Don't attempt to do something that you may not be familiar with. Stay within your own department and only give help if it is asked for. This may sound selfish, but there are good reasons for doing this. On most shoots there are specific guidelines regarding each department and the job responsibilities within that department.

Each day you will be given a call time, which is the time that you should be on the set ready to work, not the time that you are to arrive at work. I recommend arriving at work anywhere from fifteen minutes to one half-hour before the call time. Showing up a little early shows your interest and desire to do a good job. If you will be traveling to an area that you are not familiar with, be sure to look at any maps and call sheets the night before so that you have an approximate idea where you are going and what time you should leave home so that you get there on time.

When traveling to a distant location for a job, be sure that you take certain personal items with you. You should have an extra change of clothes and various personal hygiene items on the camera truck in case of emergency. Remember you may be away from home for an extended period, so be prepared.

Many of your jobs will come from recommendations from other crew members, especially from people within your own department. Work hard, do a good job, and always be willing to give a little extra in the performance of your job. Whenever possible, make your superior look good. By doing this you will have more job offers than you know what to do with.

Whenever a question or problem arises, it is best to follow the chain of command. Start within your department. If you are the 2nd A.C., then go to the 1st A.C. with your question or problem. If he can't help, then you should both go to the D.P. Going over someone's head will only make you look bad and could risk your possible employment on future productions.

Being safety conscious is very important on any film set. No shot is so important that you should jeopardize an actor or crew person's safety. If you have a concern, it should be brought up immediately. I have refused to do certain shots because I felt that my personal safety was in jeopardy. In most cases you will be respected for your professionalism and willingness to speak up.

Professionalism is an important aspect of the job in many different ways. If you feel that you are being treated unfairly, you should mention it immediately. As I stated earlier, you are a professional and should act accordingly. You should also be treated as a professional. A situation that I was in a few years ago illustrates this. I was hired on a production as the 2nd A.C. During the interview I was told that an overtime rate would be paid on hours worked past 12 hours per day. We were using two cameras, and as the 2nd A.C. the job sometimes required me to continue working forty-five minutes to an hour or more after most of the crew had wrapped and gone home for the day. I had to prepare the film to be sent to the lab, complete all paperwork, and get the equipment ready for the next day's shooting. At the start of the second week of filming, the production manager came to me and told me that he couldn't pay me for the overtime hours that I had put on my time card. I reminded him of the agreement regarding overtime pay and he told me that overtime was based on when the official wrap time was called for the entire crew. I explained that my job required me to work longer each day to complete the above-mentioned duties. He said he was sorry but that he could only pay overtime based on the crew wrap time. I looked at him and said, very sternly, "Fine, when you call wrap tonight, I am going home. You unload the film and complete all of the paperwork." At this point I walked back onto the set. A short time later the Producer called me into his office. I explained that we had agreed on an overtime deal during the job interview, and if he wasn't going to honor his agreement then I would leave then, and he could find someone else to do the job. He then told me to put any overtime hours on the time card and promised to honor his original agreement. By standing up for what was right, I showed my professionalism. I did not let the Production Manager or Producer force me into a situation that was unfair. From that day until the end of production, the Producer showed a much greater respect for me because of my willingness to stand up for what I felt was right.

Another part of being professional is having the right tools and equipment for the job. Many of the tools in your ditty bag may not be used regularly, but having that one special item when it is needed may be the difference between you and someone else being hired for the next job. Be sure to have at least the basic tools and accessories in order to fulfill your job responsibilities.

Wrap time is the time that filming ends and the crew packs up to go home for the day. When wrap is called by the Assistant Director, try to put everything away and leave as quickly as possible. Especially as the 2nd A.C., if you have kept up on the magazines and paperwork throughout the day, you should have minimal work to do at the wrap. Remember, the faster you wrap, the sooner you get home.

Being truthful on the time card is just as important as being truthful on the resume. During the job interview you should have worked out the deal for overtime, etc. By filling in your time card accurately and truthfully, you will also show your professionalism and will get more jobs.

As part of a film crew, there will be many times when you are filming on location, in offices, business establishments, people's homes, etc. Whenever you are on a location you should respect these people's homes, businesses, and property. Just because you are on a film crew, doesn't give you the right to act as you please. The proper attitude and behavior applies as much to location work as it does when you are working on a stage or in a studio.

When filming in any situation, whether it is on location or in a studio, there are certain commonsense guidelines that you should be aware of. You should avoid any type of sexual, racial, political, or religious comments which could offend others. Avoid the use of profanity as much as possible. The use of drugs or alcohol before, during, or after work is not recommended. Avoid negative comments or opinions about other production companies, rental companies, equipment, or crew members. The production community is very small, and any bad things you may say now will come back to haunt you sooner or later. Any of the above types of behavior only shows a non-professional attitude.

AFTER THE JOB

Many of the items discussed in the previous section can also be applied here. One of the main things that you should do after the job is to stay in contact with certain crew members. This industry relies heavily on networking and word of mouth. Many of your jobs will come from the recommendations of other crew people that you have worked with in the past. By staying in touch, you will keep your name fresh in their minds, and when the next job comes up they may call you.

It is especially important to stay in touch with the camera crew members that you have worked with in the past. Periodically call them and let them know that you are available for any future projects. Often a Camera Assistant may get a job call, but because of a conflict with another job, will have to turn it down. If you stay in touch with other camera crew people, you may be recommended for a job that another assistant turned down. Also, if you must turn down a job because of a conflicting job, be sure to recommend a fellow assistant for the position. And remember, whenever you are forced to turn down a job, be sure to tell the production company that it is because of a conflict with another job. A production company will be more inclined to call you again if they know that you work steady. Steady work is usually an indication that you are good at what you do.

After a job, call the D.P. to thank him or her for having you as part of the crew. Let the D.P. know that you would like to work with him or her again, and ask if you may call from time to time to keep in touch. Sometimes the first person to be hired for a job is the most recent person that the D.P. talked with. Let's hope that person is you. Good luck and have fun.

Appendix A

Film Stock

The two manufacturers of professional motion picture film stock are Eastman Kodak and Fuji. Today there are a large number of film stocks available for both 16mm and 35mm professional cinematography. These film stocks are available in color and black & white, in negative and reversal. Emulsions are available in slow-, medium-, and high-speed exposure index (E.I.) ratings. Some of the film stocks are balanced for shooting in tungsten light and some are balanced for shooting in daylight. Table A.1 contains a listing of all current film stocks available from Eastman Kodak and Fuji at publication time.

Rolls of film come in various lengths because of the different camera and magazine sizes in use today. Table A.2 lists the standard packaging sizes for motion picture film. When choosing a film stock, check with the manufacturer or distributor to be sure that it is available in the size and type that will suit your filmmaking needs.

Table A.1 Professional Motion Picture Film Stock

FILM STOCK	16mm or 35mm	COLOR BALANCE	EI or ASA TUNGSTEN	EI or ASA DAYLIGHT
EASTMAN KODAK COLOR NEGATIVE				
7245	16mm	Daylight	12 w / 80A	50
7248	16mm	Tungsten	100	64 w / 85
7287	16mm	Tungsten	200	125 w / 85
7293	16mm	Tungsten	200	125 w / 85
7297	16mm	Daylight	64 w /80A	250
7298	16mm	Tungsten	500	320 w / 85
5245	35mm	Daylight	12 w / 80A	50
5247	35mm	Tungsten	125	80 w / 85
5248	35mm	Tungsten	100	64 w / 85
5287	35mm	Tungsten	200	125 w / 85
5293	35mm	Tungsten	200	125 w / 85
5296	35mm	Tungsten	500	320 w / 85
5297	35mm	Daylight	64 w /80A	250
5298	35mm	Tungsten	500	320 w / 85
EASTMAN KODAK COLOR REVERSAL				
7239	16mm	Daylight	40 w / 80A	160
7240	16mm	Tungsten	125	80 w / 85
7250	16mm	Tungsten	400	250 w / 85
7251	16mm	Daylight	100 w / 80A	400
EASTMAN KODAK BLACK & WHITE NEGATIVE				
7222	16mm	B & W	200	250
7231	16mm	B & W	64	80
5222	35mm	B & W	200	250
5231	35mm	B & W	64	80
EASTMAN KODAK BLACK & WHITE REVERSAL				
7276	16mm	B & W	40	50
7278	16mm	B & W	160	200
FUJI COLOR NEGATIVE				
8621	16mm	Daylight	16 w / 80A	64
8631	16mm	Tungsten	125	80 w / 85
8651	16mm	Tungsten	250	160 w / 85
8661	16mm	Daylight	64 w / 80A	250
8671	16mm	Tungsten	500	320 w / 85
8521	35mm	Daylight	16 w / 80A	64
8531	35mm	Tungsten	125	80 w / 85
8551	35mm	Tungsten	250	160 w / 85
8561	35mm	Daylight	64 w / 80A	250
8571	35mm	Tungsten	500	320 w / 85

Table A.2 Film Packaging Sizes

16mm	35mm
100 ft. on Daylight Spool	100 ft. on Daylight Spool
200 ft. on Daylight Spool	200 ft. on 2" Plastic Core
400 ft. on Daylight Spool	400 ft. on 2" Plastic Core
400 ft. on 2" Plastic Core	1000 ft. on 2" Plastic Core
1200 ft. on 3" Plastic Core	2000 ft. on 3" Plastic Core

Appendix B

Equipment

As a Camera Assistant you need to have a working knowledge of all of the equipment that you use on a daily basis. This section contains listings of the various cameras, filters, heads, and tripods that you should be familiar with. When working with a piece of equipment for the first time, it is a good idea to check it out with the rental house so that you are familiar with how it works. Most rental houses are willing to help and show you any piece of equipment that you are not familiar with.

Don't walk into a rental house and expect them to drop everything to show you a particular piece of equipment. Call them ahead of time. Ask when would be a convenient time for you to come in so that they can show it to you. This is especially important if the equipment is being rented from that rental house. If you establish a good relationship with them, they will be more willing to help you out in the future.

CAMERAS

This is a basic list of the most commonly used 16mm and 35mm cameras in use today. I have also included the two 65mm cameras currently available. If you are not familiar with a specific camera, ask the rental house personnel to show it to you and explain how it works. Chapter 6 contains simple illustrations and threading diagrams of most of these listed cameras and their magazines.

16mm Cameras

Aaton XTR-Plus
Aaton XTR-Prod
Arriflex 16BL
Arriflex 16M
Arriflex 16S/SB
Arriflex 16SR1 & 16SR2
Arriflex 16SR3
Bolex H-16
Canon Scoopic 16MS
Cinema Products CP16R

Eclair ACL
Mitchell Professional
Photo-Sonics 1PL (high speed)
Eclair NPR
Panavision Panaflex 16

35mm Cameras

Aaton 35
Arriflex 435
Arriflex 35-3
Bell & Howell Eyemo
Image 300 (high speed)
Mitchell/Fries 35 R
Mitchell NC, NCR
Mitchell S 35 R/MKII, S 35 RB
Mitchell High Speed
Moviecam Super America
Panavision Golden
Panavision Panaflex-X
Panavision PSR
Photo-Sonics 4B/4C (high speed)
Photo-Sonics 4ML
Arriflex 535
Arriflex 35BL3, 35BL4
Arriflex 2-C
Cinema Products XR 35
Leonetti Ultracam
Mitchell/Fries 35 R3
Mitchell BNC, BNCR
Mitchell Standard
Moviecam Compact
Panavision Platinum
Panavision Golden G-II
Panavision Panastar I & II (high speed)
Panavision Super PSR
Photo-Sonics 4E/ER (high speed)

65mm Cameras

Arriflex 765
Panavision System 65

CAMERA FILTERS

There are a wide variety of filters available for motion picture cameras. Each filter has its own specific effect and is chosen based on the D.P.'s preference. Filters are available in many different sizes. Some filters are available in varying densities, with the lower numbers being lighter density and the higher numbers being heavier density. The most common camera filters for motion picture photography are manufactured by Tiffen, Harrison & Harrison, Mitchell, Wilson Film Services, and Fries Engineering.

The following are the various sizes that most of the filters are available in and also the most commonly used filters used in motion picture photography. The numbers following some of the filters indicate the densities that they are available in. The smaller numbers indicate a very light effect and the larger numbers indicate a heavier effect. For example a diffusion 1/8 would have a lesser effect on the image than a diffusion 2.

Filter Sizes

40.5mm round	48mm round
Series 9 or 3 1/2″ round	4 1/2″ round
138mm or 5 1/2″ round	3″ × 3″ square
4″ × 4″ square	4″ × 5.65″ Panavision
5″ × 6″	6.6″ × 6.6″ square

Filters

85, 85N3, 85N6, 85N9, 85B, 85C, 85 POLA

ND3, ND6, ND9, ND12

80A, 80B, 80C

81A, 81B, 81C, 81EF

82, 82A, 82B, 82C

812

Polarizing

LLD

Enhancer

Clear: optical flat

Red, green, glue, cyan, yellow, magenta, pink, sunset, sepia, chocolate, tobacco

ND3 grad, ND6 grad, ND9 grad, ND12 grad

Color grads: red, green, blue, cyan, yellow, magenta, pink, sunset, chocolate, sepia, tobacco, cranberry, plum, tangerine, straw, grape

Color compensating: blue, cyan, green, magenta, red, yellow

Coral: 1/8, 1/4, 1/2, 1, 2, 3, 4, 5, 6, 7, 8

Coral grad: 1/8, 1/4, 1/2, 1, 2, 3, 4, 5

Fog: 1/8, 1/4, 1/2, 1, 2, 3, 4, 5

Double fog: 1/8, 1/4, 1/2, 1, 2, 3, 4, 5

Low contrast: 1/8, 1/4, 1/2, 1, 2, 3, 4, 5

Ultra contrast: 1/8, 1/4, 1/2, 1, 2, 3, 4, 5

Soft contrast: 1, 2, 3, 4, 5

Soft FX: 1, 2, 3, 4, 5

Warm soft FX: 1/2, 1, 2, 3, 4, 5

Softnet: black, white, red, skintone—1, 2, 3, 4, 5

Diopters: +1/2, +1, +1 1/2, + 2, +3

Split diopters: +1/2, +1, +1 1/2, +2, +3

Star: 4 pt., 6 pt., 8 pt. (available in 1mm, 2mm, 3mm, or 4mm grid pattern.)

Fluorescent light correction: FLD (daylight), FLB (tungsten)

Black dot texture screens

Day for night effects

Mitchell Diffusion: A, B, C, D, E

Supafrost: 00, 00+, 0, 0+, 1, 1+

Black Supafrost: 00, 00+, 0, 0+, 1

Diffusion: 1/2, 1, 1 1/2, 2, 2 1/2, 3

White Pro Mist: 1/8, 1/4, 1/2, 1, 2, 3, 4, 5

Black Pro Mist: 1/8, 1/4, 1/2, 1, 2, 3, 4, 5

Warm Pro Mist: 1/8, 1/4, 1/2, 1, 2

Black Net: 1, 2, 3, 4, 5

White Net: 1, 2, 3, 4, 5

Filters for Black & White Cinematography

#6 Yellow
#9 Yellow
#12 Yellow
#15 Deep Yellow
#21 Orange
#25 Red
#8 Yellow
#11 Green
#13 Green
#16 Orange
#23A Light Red
#25A Red

#29 Dark Red
#47 Blue
#47B Dark Blue
#56 Light Green
#58 Green
#61 Dark Green

HEADS AND TRIPODS

In addition to having a working knowledge of the cameras and filters in use today, a Camera Assistant should know the various heads and tripods on which the cameras may be mounted. The following are the most commonly used heads and tripods:

Fluid Heads

Cartoni C-40
Cartoni Dutch Head
O'Connor 150 B
O'Connor 100C HD
O'Connor 100 C
O'Connor 50 D
O'Connor Ultimate
Ronford Baker Fluid 30
Ronford Baker 15 S
Ronford Baker Fluid 15
Ronford Baker Fluid 7 MK III
Sachtler Studio 7 + 7
Sachtler Standard 7 + 7
Sachtler Panorama 7 + 7
Sachtler Horizon 7 + 7
Sachtler Video 20
Sachtler Video 25
Sachtler Studio 80
Sachtler Dutch Head
Weaver-Steadman

Geared Heads

Arriflex Arrihead
Cinema Products Mini-Worrall
Cinema Products Worrall
Mitchell Lightweight
NCE/Ultrascope MK III
Panavision Panahead
Panavision Super Panahead
Technovision Technohead MK III

Tripods

Bazooka

O'Connor wooden tripod with Mitchell flat top casting—standard and baby

O'Connor wooden tripod with ball top casting—standard and baby

Panavision Panapod with Mitchell flat top casting—standard and baby

Ronford aluminum tripod with Mitchell flat top casting—standard and baby

Ronford aluminum tripod with ball top casting—standard and baby

Sachtler aluminum tripod with ball top casting—standard and baby

Appendix C

Checklists

Included in this section are many different checklists that you will need when you begin working as a Camera Assistant. By using these checklists, you will be sure that you have all equipment and supplies needed for your shoot. Table C.1 contains a camera equipment checklist, and Table C.2 contains a filter checklist. These will help when you do the camera prep. Table C.3 is the expendables checklist, which will ensure that you have all the expendables needed in order to complete the job.

Table C.1 Camera Equipment Checklist

X	CAMERAS		X	VIDEO ASSIST		X	PRIME LENSES	
	Aaton 35			Arri 35-3 Video Door			3.5 mm	27 mm
	Aaton XTR Plus			B & W Video Tap			4 mm	28 mm
	Aaton XTR Prod			Color Video Tap			5.9 mm	30 mm
	Arriflex 535			Wireless Transmitter			8 mm	32 mm
	Arriflex 435			4" Monitor			9.5 mm	35 mm
	Arriflex 35 BL			9" Monitor			9.8 mm	40 mm
	Arriflex 35-3			12" Monitor			10 mm	50 mm
	Arriflex 2-C			19" Monitor			12 mm	55 mm
	Arriflex 16 SR2			Video 8 Combo			14 mm	60 mm
	Arriflex 16 SR3			Watchman			14.5 mm	65 mm
	Arriflex 16 SR2 HS			VHS Deck			16 mm	75 mm
	Arriflex 16 SR3 HS			25' Coaxial			17 mm	85 mm
	Bell & Howell Eyemo			50' Coaxial			17.5 mm	100 mm
	Leonetti Ultracam			Adapters, Barrel Connectors			18 mm	125 mm
	Moviecam Super America						20 mm	135 mm
	Moviecam Compact						21 mm	150 mm
	Panavision Platinum		X	HAND HELD			24 mm	180 mm
	Panavision G2			Right Hand Grip			25 mm	
	Panavision Panaflex-X			Left Hand Grip				
	Panavision Panastar			Shoulder Pad				
	Panavision 16			Follow Focus				
				Clamp-on Matte Box		X	ZOOM LENSES	
				Arri 35-3 Hand Held Door			7 - 56	18 - 100
							8 - 64	18.5 - 55.5
X	GROUND GLASS						9 - 50	20 - 60
	1.33	1.66					9.5 - 57	20 - 100
	1.85	2.35	X	TELEPHOTO LENSES			10.4 - 52	20 - 120
	TV	TV/1.85		200 mm	800 mm		10 - 30	20 - 125
				300 mm	1000 mm		10 - 100	23 - 460
				400 mm	1200 mm		10 - 150	24 - 275
X	MAGAZINES			500 mm	2000 mm		11 - 110	25 - 250
	200	250		600 mm			11.5 - 138	25 - 625
	400	500					12 - 120	27 - 68
	1000	1200					12 - 240	35 - 140
	2000	Reverse	X	MACRO LENSES			14 - 70	40 - 200
				16 mm	60 mm		16 - 44	48 - 550
	Arri 35-3 Shoulder			24 mm	75 mm		17 - 75	50 - 500
	Arri 35-3 Steadicam			32 mm	90 mm		17.5 - 75	135 - 420
	535 Steadicam			40 mm	100 mm		17 - 102	150 - 600
	435 Shoulder			50 mm	200 mm		18 - 90	
						X	VARIABLE PRIMES	
							16 - 30	55 - 105
							29 - 60	

Table C.1 *(continued)*

X	ACCESSORIES	X	HEADS	X	LENS ACCESSORIES
	Eyepiece Extension		Arrihead		Short Iris Rods
	Mid Range Eyepiece		Panahead		Medium Iris Rods
	Eyepiece Leveler		Super Panahead		Long Iris Rods
	Eyepiece Heater		Mitchell Gear Head		Zoom Bridge Support
	Eyepiece Heater Cable		Mini Worral		Matte Box Adapter Rings
	Sliding Balance Plate		Worral Gear Head		Matte Box Bellows
	Auxiliary Carry Handle		Sachtler 7 + 7		Rubber Donuts
	Studio Follow Focus		Sachtler Video 80		Lens Shade
	Mini Follow Focus		O'Connor 100		Microforce Motor
	Remote Follow Focus		O'Connor Ultimate		Microforce Zoom Control
	Focus Gears		Ronford 7		Panavision Zoom Control
	Focus Whip - 6", 12"		Ronford S-15		Panavision Zoom Holder
	Speed Crank		Weaver Steadman		Zoom Power Cables
	4 x 4 Matte Box		Cartoni		Microforce Extension Cable
	4 x 5.65 Matte Box		Vitesse		Low Angle Prism
	5 x 6 Matte Box		Dutch Head		Century Periscope
	6.6 x 6.6 Matte Box		Tilt Plate		1.4X Extender
	4 x 4 Filter Trays		Rocker Plate		2X Extender
	4 x 5.65 Filter Trays				Aspheron for 9.5 & 12
	6.6 x 6.6 Filter Trays				Mutar for 10-100
	Filters-See Filters Check List	X	**SUPPORT**		Panavision/Frazier Lens System
	Hard Mattes		Standard Tripod		
	Eyebrow		Baby Tripod		
	4 1/2" Clamp-On Shade		Bazooka		
	138mm Clamp-On Shade		High Hat	X	**BATTERIES**
	Extra Filter Retaining Rings		Low Hat		Blocks
	Remote Switch		Spreader		Belts
	Lens Light				Chargers
	HMI Speed Control				On-Board
	Precision Speed Control	X	**MISCELLANEOUS**		On-Board Chargers
	Intervalometer		400' Film Cans		Arri SR On-Board Adapter
	Capping Shutter		1000' Film Cans		Power Cables
	Film/Video Synchronizer		400' Black Bags		
	Junction Box		1000' Black Bags		
	Obie Light		Film Cores		
	Rain Cover		Camera Reports	X	**OTHER**
	Camera Barney		Changing Bag or Tent		
	Magazine Barney		Slate		
	Utility Base Plate		Expendables - See Check List		
	Rain Deflector				
	Director's Finder				
	Microforce Handle for Sachtler				

Table C.2 Filters Checklist

MULTI-PURPOSE FILTERS	4 1/2	138	4X4	4X5	6X6		
85, 85N3, 85N6, 85N9							
ND3, ND6, ND9							
POLARIZER							
80A							
81EF							
LLD							
85 POLA							
OPTICAL FLAT							
ENHANCER							
ND GRADS - SOFT EDGE ND3, ND6, ND9							
ND GRADS - HARD EDGE ND3, ND6, ND9							
CORAL 1/8, 1/4, 1/2, 1, 2, 3							
CORAL GRADS 1/8, 1/4, 1/2, 1, 2, 3							
DIFFUSION 1/8,1/4, 1/2, 1, 2, 3							
BLK PROMIST 1/8, 1/4, 1/2, 1, 2							
WHT PROMIST 1/8, 1/4, 1/2, 1, 2							
WARM PROMIST 1/8, 1/4, 1/2, 1, 2							
MITCHELL A, B, C, D, E							
BLACK NET 1, 2, 3, 4, 5							
WHITE NET 1, 2, 3, 4, 5							
BLK SOFTNET 1, 2, 3, 4, 5							
WHT SOFTNET 1, 2, 3, 4, 5							
RED SOFTNET 1, 2, 3, 4, 5							
BLACK DOT 1, 2, 3, 4, 5							
LOW CON 1/8, 1/4, 1/2, 1, 2, 3							
SOFT CON 1/8, 1/4, 1/2, 1, 2, 3							
ULTRA CON 1/8, 1/4, 1/2, 1, 2, 3							
SOFT F/X 1/2, 1, 2, 3, 4, 5							
FOG 1/8, 1/4, 1/2, 1, 2, 3							
DBL FOG 1/8, 1/4, 1/2, 1, 2, 3							
DIOPTER +1/2, +1, +1 1/2, +2, +3							
SPLIT DIOPTER +1/2, +1, +2, +3							
FLUORESCENT FLB							
FLUORESCENT FLD							
STAR 1, 2, 3, 4 4 pt. 6 pt. 8 pt.							
DAY FOR NIGHT							
COLOR FILTERS							
RED 1, 2, 3, 4, 5							
BLUE 1, 2, 3, 4, 5							
GREEN 1, 2, 3, 4, 5							
YELLOW 1, 2, 3, 4, 5							
CYAN 1, 2, 3, 4, 5							
MAGENTA 1, 2, 3, 4, 5							
CHOCOLATE 1, 2, 3							
SEPIA 1, 2, 3							
TOBACCO 1, 2, 3							

Table C.2 (*continued*)

COLOR GRAD FILTERS	4 1/2	138	4X4	4X5	6X6		
RED 1, 2, 3, 4, 5							
BLUE 1, 2, 3, 4, 5							
COOL BLUE 1, 2, 3, 4, 5							
TROPIC BLUE 1, 2, 3, 4, 5							
GREEN 1, 2, 3, 4, 5							
YELLOW 1, 2, 3, 4, 5							
CYAN 1, 2, 3, 4, 5							
MAGENTA 1, 2, 3, 4, 5							
PINK 1, 2, 3, 4, 5							
SUNSET 1, 2, 3							
CHOCOLATE 1, 2, 3							
SEPIA 1, 2, 3							
TOBACCO 1, 2, 3							
CRANBERRY 1, 2, 3							
TANGERINE 1, 2, 3							
STRAW 1, 2, 3							
PLUM 1, 2, 3							
GRAPE 1, 2, 3							
FILTERS FOR BLACK & WHITE							
#6 YELLOW							
#8 YELLOW							
#9 YELLOW							
#11 GREEN							
#12 YELLOW							
#13 GREEN							
#15 DEEP YELLOW							
#16 ORANGE							
#21 ORANGE							
#23A LIGHT RED							
#25 RED							
#25A RED							
#29 DARK RED							
#47 BLUE							
#47B DARK BLUE							
#56 LIGHT GREEN							
#58 GREEN							
#61 DARK GREEN							

Table C.3 Expendables Checklist

	1" Camera Tape - Black		Fine Point Sharpies - Black
	1" Camera Tape - White		Fine Point Sharpies - Red
	1" Camera Tape - Red		Fine Point Sharpies - Blue
	1" Camera Tape - Yellow		Extra Fine Point Sharpies - Black
	1" Camera Tape - Blue		Extra Fine Point Sharpies - Red
	1" Paper Tape - Red		Extra Fine Point Sharpies - Blue
	1" Paper Tape - Green		Medium Point Ball Point Pens - Black
	1" Paper Tape - Blue		Fine Point Ball Point Pens - Black
	1" Paper Tape - Orange		Wide Marker - Black
	1" Paper Tape - Yellow		Wide Marker - Red
	1" Paper Tape - White		Wide Marker - Blue
	1" Paper Tape - Black		Stabilo Grease Pencil - White
	1/2" Paper Tape - Red		Stabilo Grease Pencil - Red
	1/2" Paper Tape - Green		Stabilo Grease Pencil - Yellow
	1/2" Paper Tape - Blue		Vis-A-Vis Erasable Felt Marker - Black
	1/2" Paper Tape - Orange		Vis-A-Vis Erasable Felt Marker - Red
	1/2" Paper Tape - Yellow		Vis-A-Vis Erasable Felt Marker - Blue
	1/2" Paper Tape - White		Vis-A-Vis Erasable Felt Marker - Green
	1/2" Paper Tape - Black		Chalk - Box
	2" Gaffer Tape - Gray		Yellow Legal Pad - 8 1/2" x 11"
	2" Gaffer Tape - Black		Kodak Wratten Gel - 85
	2" Gaffer Tape - White		Kodak Wratten Gel - 85 N3
	2" Paper Tape - Black		Kodak Wratten Gel - 85 N6
	1/8" Chart Tape White Yellow		Kodak Wratten Gel - 85 N9
	1/4" Chart Tape White Yellow		Kodak Wratten Gel - ND 3
	Rosco Lens Tissue		Kodak Wratten Gel - ND 6
	Rosco Lens Fluid		Kodak Wratten Gel - ND 9
	Chamois Eyepiece Covers - Large		Kodak Wratten Gel - 80 A
	Chamois Eyepiece Covers - Small		Orangewood Sticks
	Write On/Wipe Off Slate Marker - Black		Camera Wedges
	Write On/Wipe Off Slate Marker - Blue		Color Chart
	Write On/Wipe Off Slate Marker - Red		Gray Scale
	Powder Puffs for Slate		Insert Slate
	1/2" Stick On Letters & Numbers - Black		Sync Slate
	1/2" Stick On Letters & Numbers - Red		AAA Alkaline Batteries
	1/2" Stick On Letters & Numbers - Blue		AA Alkaline Batteries
	3/4" Stick On Letters & Numbers - Black		C Alkaline Batteries
	3/4" Stick On Letters & Numbers - Red		D Alkaline Batteries
	3/4" Stick On Letters & Numbers - Blue		9-Volt Alkaline Batteries
	Dust-Off		Light Meter Batteries #
	Dust-Off Nozzle		Mag Lite Replacement Bulbs
	Dust-Off Plus		Paper Towels
	Dust-Off Plus Nozzle		Box of Rags
	Cotton Swabs (Q-Tips)		Silicone Spray
	Foamtip Swabs		Lighter Fluid
	Velcro - 1"		Spray Cleaner (409, Simple Green)
	Velcro - 2"		Large Plastic Trash Bags
	Small Kimwipes		
	Large Kimwipes		

Appendix D

Tools and Accessories

One of the main requirements of a professional Camera Assistant is to have a basic set of tools, accessories, and expendables to help you to do the job properly. Some of the tools are basic everyday items that you may need to perform minor repairs on the camera. Others are specialized pieces of equipment that are unique to the film industry. Some of the items in your tool kit or ditty bag should include many of the items on the camera department expendable list. Many assistants also have certain specialty items in their ditty bag based on personal preference. The following list has a basic set of tools and accessories, and a brief description of what each is used for. Also listed are the expendables and a description of what each is used for.

Tools and Accessories

Sync slate: to slate sync sound shots

Small sync slate: for slating close-up sync sound shots

Insert slate: to slate silent or insert shots

Large clapper sticks: for slating multiple camera shots

50′ cloth tape measure: to measure focus distances to the actors

25′ metal tape measure: for measuring to points on the set, and for placing marks along the floor (not recommended for measuring to actors because of danger of injury from the metal edge)

Changing bag or changing tent: for loading and unloading film when no darkroom is available

Scissors: for cutting film along the perforations so that it has a smooth edge for threading

Magnifier with light: for checking the gate

Small flashlight: for checking the gate

Large flashlight: useful on dark sets or locations to see your way

French flag with arm: to flag lights that are flaring the lens

6″ focus whip: an extension for the follow focus mechanism to allow the first assistant to better follow focus during shooting

Color chart and gray scale: to photograph at the head of the first roll of film for the lab to use in timing

Tweezers: for removing pieces of film from inside the camera body

Blower bulb syringe: to clean dirt and dust off the lens and filters

Bubble level: to ensure that the camera is level on uneven surfaces

Small pocket level: to ensure that the camera is level on uneven surfaces

Ground glass puller: to aid in removing the ground glass, especially in some Arriflex cameras

Camera oil (Mitchell, Panavision, Arriflex): to be used to oil the camera at proper intervals

Camera silicone lubricant: for lubricating camera pull down claw when necessary

Electrical adapters: for use when plugging in camera batteries for charging

Cube taps: same as electrical adapters

Depth-of-field charts or calculator: to determine your depth of field for a particular shot

Set of jeweler's screwdrivers: for minor camera repairs

Slotted screwdriver—1/8″, 3/16″, 1/4″, 5/16″: for minor camera repairs

Phillips screwdrivers: #1, #2: for minor camera repairs

Magnetic screwdriver: for making basic repairs on the camera

Allen wrenches—metric and American: for minor camera repairs

Adjustable wrenches: for minor camera repairs

Vise grips: for minor camera repairs

Pliers: for minor camera repairs

Needlenose pliers: for minor camera repairs

Razor knife: used for cutting gels or anything else that needs cutting

Wire cutters: for minor camera and power cable repairs

Leatherman tool: for minor repairs on the camera

Swiss Army knife: various uses, including repairs to camera

3/8″-16 bolts—short and long: used for mounting camera body to various surfaces

Rubber T-marks: for marking actors, especially outside where tape will not stick

Space blanket: for covering the camera to protect it from weather or heat

Velcro cable ties: to keep power cables coiled up

Engraved filter tags: for identifying filters placed on the camera or in the matte box

Assorted video connectors and cables: for connecting the video tap to monitors or video recorders

Empty filter pouches (various sizes): for storing and protecting filters

1″ brush: for dusting off the camera and equipment

Camera fuses: for use in case the fuse in the camera blows

Oil dropper or syringe: used for oiling the camera

Duvatene (black cloth): various uses, including covering the camera if it is reflected in glass or a mirror

Convex mirror: placed in front of the lens to check for lens flares

Extra power cables: used as spare in case power cables become worn, frayed, or damaged

Dental mirror: used when adjusting the shutter and synching the camera to a video monitor or projector screen

Small C-clamps: used to secure the camera when mounting it to unsteady surfaces

Small grip clamps: used for securing space blanket or duvatene to the camera

Front box: for storage of basic accessories used daily by the 1st A.C., and mounted on the head so that these accessories are readily available

ASC Manual: used as a reference

Professional Cameraman's Handbook: used as a reference

The Camera Assistant's Manual: used as a reference

Camera Terms and Concepts: used as a reference

Various camera instruction manuals: used as a reference

Expendables

1″ camera tape—black, white, red, yellow, blue: for wrapping cans and making labels and marks

2″ gaffer tape—black, gray: for labeling equipment and securing various items during production

1/2″ or 1″ paper tape—red, blue, green, yellow, white, orange: for marking actors

1/8″ or 1/4″ chart tape—white, yellow: for marking focus distances on the lens

Lens tissue: to clean lenses and filters along with lens cleaner

Lens cleaner: to clean lenses and filters along with lens tissue

Orangewood sticks: to clean emulsion buildup from the gate

Eyepiece covers: for the camera operator's comfort when looking through eyepiece

Permanent felt tip markers (Sharpies)—red, black, blue: for making labels

Wide tip permanent felt tip markers (Magic Marker)—red, black, blue: for labeling equipment cases

Ballpoint pens: to fill out camera reports, inventory forms, time sheets, etc.

Grease pencils (Stabilo)—white, yellow, red: for marking the film and making focus marks on the lens

Vis-A-Vis erasable felt markers—black, red, green, blue: for marking focus distances on follow focus marking disc or with chart tape when making marks directly on lens

Slate marker (Write On/Wipe Off): for writing information on the slate

Makeup-type powder puff: for erasing information written on slate with slate markers

Chalk: to mark the position of the camera, dolly, or actors

Camera wedges: to aid in leveling the camera when it is placed on a high hat or any uneven surface

Stick-on letters—black, red, blue: for placing production information on the slate

Compressed air with nozzle (Dust Off): same as blower bulb syringe and also to clean the inside of the camera body

Cotton swabs or foam-tip swabs (Q-Tips): used for removing excess oil from camera movement

Kimwipes—large and small: used for small cleaning jobs on camera and equipment (not to be used on camera lenses, but may be used on filters along with cleaning solution)

Silicone spray: to lubricate parts that may be sticking (not to be used on the interior of the camera body)

Lighter fluid: to remove glue or sticky residue from tape or stick-on letters

Kodak Wratten gels (85, 85N3, 85N6, 85N9, ND3, ND6, ND9, 80A): for placing filters behind the lens, as in the Panavision cameras, or when a glass filter in front of the lens is not available

Spare batteries (AAA, AA, C, D, 9-volt): for powering mini mag lights or magnifiers

Spare Mini Mag Lite bulbs: replacement bulbs for the small Mini Mag Lite flashlight used by most assistants

Trash bags: for disposing of any trash and also for covering the camera and equipment in the rain

Spray cleaner: for cleaning camera equipment and cases during production and also at wrap

Rags, paper towels: used along with spray cleaner to clean equipment and cases

Miscellaneous Items

Spare cores: for threading film in the magazines

Extra film cans and black bags (400′ & 1000′): for canning out exposed film and short ends

Camera reports: for keeping a record of all scenes shot for each roll of film

Film inventory forms: for keeping a record of all film shot and received each day

Rental catalogs: used as a reference when ordering additional equipment

Along with the above-listed items, you should have some type of hard-side or soft-side case or cases to store and transport all of these items. These tools and accessories are important to the performance of your job, so it is a good idea to protect them and keep them in good condition when not being used. The type of case you use is a matter of personal preference. This listing is subject to change depending on the individual needs of each camera assistant. You may find a particular item that helps you to perform the job better. There is no right or wrong as to what you should have in your bag, only what you need to do the job.

In addition to the above-listed items many assistants have some type of pouch that they wear on their belt to keep specific items available at all times. These are items that are needed regularly during shooting and include the following: permanent felt tip markers, ballpoint pens, grease pencil, lens tissue, lens cleaner, slate marker with powder puff attached, small flashlight, magnifier with light, and depth-of-field calculator. Many assistants will also make a small loop of rope on which to keep a roll of black and white camera tape with them in case they need to make marks or labels of any kind. And finally, since you will be working with a great number of equipment cases, not including your own equipment, you should have some type of four-wheel dolly to assist in moving and transporting the equipment cases on the set. The most common type is the magliner, which is illustrated and discussed in Chapter 4.

As you gain more experience as an assistant, you will probably find other tools and accessories that you will keep in the ditty bag or AKS case. This list is meant only as a guideline for people starting out who want to acquire the basics for doing the job.

Appendix E

Tables

Included in this section are many tables that a Camera Assistant may refer to for a variety of information. These include footage tables, hyperfocal distances, f-stop compensation for changes in frames per second, f-stop compensation for various filters used, footage to time conversions, time to footage conversions, and many more.

INTERMEDIATE F-STOP VALUES

When the Director of Photography (D.P.) measures the light, the f-stop reading that he gets will not always be exactly on one of the f-stop values that I mentioned in Chapter 1. Very often the value of the light measurement will fall between two f-stop numbers. Table E.1 gives the intermediate f-stop values between each successive pair of f-stop numbers. For example, the f-stop value that is halfway between f 4 and f 5.6 is f 4.8.

F-STOP COMPENSATION WHEN USING FILTERS

During the course of shooting, the D.P. will ask that various filters be placed on the camera in order to achieve a specific effect or to correct for the color temperature. Many of the filters used will reduce the amount of light that reaches the film. Remember, that if a filter requires an f-stop compensation, the amount of compensation listed, refers to how much you should open up the lens aperture. Table E.2 lists the f-stop compensation for various filters that you may be using. For example, when using an 85 filter, you must open up your f-stop $^{2}/_{3}$ of a stop.

Table E.1 Intermediate F-Stop Values for 1/4, 1/3, 1/2, 2/3, and 3/4 Stops

FULL STOP	1/4 STOP	1/3 STOP	1/2 STOP	2/3 STOP	3/4 STOP	FULL STOP
1	1.1	1.1	1.2	1.3	1.3	**1.4**
1.4	1.5	1.6	1.7	1.8	1.9	**2**
2	2.1	2.2	2.4	2.5	2.6	**2.8**
2.8	3.1	3.2	3.3	3.6	3.7	**4**
4	4.4	4.5	4.8	5	5.2	**5.6**
5.6	6.2	6.4	6.7	7.1	7.3	**8**
8	8.7	9	9.5	10	10.5	**11**
11	12.3	12.7	13.5	14.3	14.6	**16**
16	17.4	18	19	20	21	**22**
22	24.6	25.5	27	28.6	29.2	**32**
32	34.9	35.9	38	40.3	41.5	**45**

F-STOP COMPENSATION WHEN USING FILTERS FOR BLACK & WHITE

When shooting black & white film, the D.P. will use specific filters to change the way that specific colors appear in black and white. All colors are reproduced as a certain shade of gray. By using a filter, the D.P. can alter or change how light or dark the shade of gray is for a particular color. Table E.3 lists the f-stop compensation for filters used in black & white photography. For example, when using a #12 yellow filter, you must open up your f-stop 1 stop when shooting in daylight, and 2/3 of a stop when shooting in tungsten light.

F-STOP COMPENSATION FOR CHANGES IN FRAMES PER SECOND

When you change the speed that the film travels through the camera (f.p.s.), you are changing how long each frame is exposed to light. If you run the camera at a higher speed, each frame is exposed to light for less time, and if you run the camera at a slower speed, each frame is exposed to light for a longer time. Table E.4 lists the f-stop compensation for various changes in frames per second. For example, if you change the camera speed to 60 f.p.s., you must open your f-stop 1 1/4 stops.

Table E.2 F-Stop Compensation for Various Filters

FILTER	COMPENSATION
85	2/3 STOP
85 N3	1 2/3 STOPS
85 N6	2 2/3 STOPS
85 N9	3 2/3 STOPS
LLD	0
ND 3	1 STOP
ND 6	2 STOPS
ND 9	3 STOPS
80A	2 STOPS
80B	1 2/3 STOPS
80C	1 STOP
85B	2/3 STOP
85C	1/3 STOP
81EF	2/3 STOP
812	1/3 STOP
OPTICAL FLAT	0
POLARIZER	1 1/2 - 2 STOPS
ENHANCER	1/2 - 1 STOP
FLB	1 STOP
FLD	1 STOP
SOFTNET	1/3 - 2/3 STOP
BLACK DOT	1 STOP
CORAL	BASED ON DENSITY
SEPIA	BASED ON DENSITY
FOG	0
DOUBLE FOG	0
LOW CONTRAST	0
SOFT CONTRAST	0
ULTRA CONTRAST	0
PRO MIST	0
SOFT F/X	0
DIOPTER	0

Table E.3 F-Stop Compensation when Using Filters for Black & White

FILTER	COMPENSATION	
	DAYLIGHT	TUNGSTEN
#6 YELLOW	2/3 STOP	2/3 STOP
#8 YELLOW	1 STOP	2/3 STOP
#9 YELLOW	1 STOP	2/3 STOP
#12 YELLOW	1 STOP	2/3 STOP
#15 DEEP YELLOW	1 2/3 STOPS	1 STOP
#16 ORANGE	1 2/3 STOPS	1 2/3 STOPS
#21 ORANGE	2 1/3 STOPS	2 STOPS
#11 GREEN	2 STOPS	1 2/3 STOPS
#13 GREEN	2 1/3 STOPS	2 STOPS
#56 LIGHT GREEN	2 2/3 STOPS	2 2/3 STOPS
#58 DARK GREEN	3 STOPS	3 STOPS
#61 DARK GREEN	3 1/3 STOPS	3 1/3 STOPS
#23A LIGHT RED	2 2/3 STOPS	1 2/3 STOPS
#25 RED	3 STOPS	2 2/3 STOPS
#29 DARK RED	4 1/3 STOPS	2 STOPS
#47 DARK BLUE	2 1/3 STOPS	3 STOPS
#47B DARK BLUE	3 STOPS	4 STOPS

F-STOP COMPENSATION FOR CHANGES IN SHUTTER ANGLE

Similar to when you change your camera speed, if you change your shutter angle, you will affect how much light strikes the film. Increasing the shutter angle will allow more light to reach the film, and decreasing the shutter angle will allow less light to reach the film. Table E.5 lists the f-stop compensation for various changes in shutter angle. For example, when you change your shutter angle to 90°, you must open your f-stop 1 stop.

Table E.4 F-Stop Compensation for Changes in Frames per Second

FRAMES PER SECOND	F-STOP COMPENSATION
6	CLOSE 2 STOPS
8	CLOSE 1 2/3 STOPS
9	CLOSE 1 1/2 STOPS
10	CLOSE 1 1/3 STOPS
12	CLOSE 1 STOP
15	CLOSE 3/4 STOP
16	CLOSE 2/3 STOP
18	CLOSE 1/2 STOP
20	CLOSE 1/3 STOP
21	CLOSE 1/4 STOP
24	0
30	OPEN 1/4 STOP
32	OPEN 1/3 STOP
36	OPEN 1/2 STOP
40	OPEN 2/3 STOP
42	OPEN 3/4 STOP
48	OPEN 1 STOP
60	OPEN 1 1/4 STOP
64	OPEN 1 1/3 STOP
72	OPEN 1 1/2 STOP
80	OPEN 1 2/3 STOP
84	OPEN 1 3/4 STOP
96	OPEN 2 STOPS
120	OPEN 2 1/4 STOPS

Table E.5 F-Stop Compensation for Changes in Shutter Angle

SHUTTER ANGLE	F-STOP COMPENSATION
200/180	Full Exposure
172.8	------
157.5	- 1/4
150.3	- 1/3
144	- 1/3
135	- 1/2
120.6	- 2/3
112.5	-3/4
90	-1
78.75	- 1 1/4
75.15	- 1/13
67.5	- 1 1/2
60.3	- 1 2/3
56.25	- 1 3/4
45	-2

HYPERFOCAL DISTANCES

Hyperfocal distance is a special case of depth of field. It is defined as the closest point in front of the lens, which will be in acceptable focus, when the lens is focused to infinity. There will be certain times during filming when you need to know the hyperfocal distance for a particular shot. Tables E.6 and E.7 list the hyperfocal distances for various focal length lenses for both 16mm and 35mm formats. All amounts are rounded to the nearest inch. For example, from Table E.7, when shooting in 35mm, with a 29mm lens and an f-stop of 5.6, your hyperfocal distance is 19 feet, 5 inches.

Table E.6 16mm Hyperfocal Distances—Circle of Confusion = .0006″

FOCAL LENGTH	f-stop 1	1.4	2	2.8	4	5.6	8	11	16	22
5.9	7'6"	5'5"	3'8"	2'8"	1'11"	1'4"	11"	8"	6"	4"
8	13'10"	9"10"	6'11"	4'11"	3'5"	2'6"	1'8"	1'2"	11"	7"
9.5	19'5"	13'11"	9'8"	6'11"	4'11"	3'6"	2'5"	1'10"	1'2"	11"
10	21'6"	15'5"	10'10"	7'8"	5'5"	3'10"	2'8"	2'	1'4"	1'
12	31'	22'1"	15'6"	11'1"	7'8"	5'6"	3'11"	2'10"	1'11"	1'5"
14	42'2"	30'1"	21'1"	15'1"	10'6"	7'6"	5'4"	3'10"	2'7"	1'11"
16	55'1"	39'5"	27'6"	19'8"	13'10"	9'10"	6'11"	5'	3'5"	2'6"
17	62'2"	44'5"	31'1"	22'2"	15'7"	11'1"	7'10"	5'8"	3'11"	2'10"
18	69'8"	49'10"	34'11"	24'11"	17'5"	12'6"	8'10"	6'4"	4'5"	3'2"
20	86'1"	61'6"	43'1"	30'10"	21'6"	15'5"	10'10"	7'10"	5'5"	3'11"
21	94'11"	67'10"	47'6"	33'11"	23'8"	16'11"	11'11"	8'7"	5'11"	4'4"
24	124'	88'7"	62'	44'4"	31'	22'1"	15'6"	11'4"	7'8"	5'7"
25	134'6"	96'1"	67'4"	48'1"	33'7"	24'	16'10"	12'2"	8'5"	6'1"
27	156'11"	112'1"	78'6"	56'	39'2"	28'	19'7"	14'4"	9'10"	7'1"
28	168'7"	120'6"	84'5"	60'4"	42'2"	30'1"	21'1"	15'3"	10'5"	7'7"
29	181'	129'4"	90'6"	64'8"	45'4"	32'4"	22'7"	16'6'	11'4"	8'2"
32	220'5"	157'6"	110'2"	78'8"	55'1"	39'4"	27'6"	20'	13'10"	10'
35	263'8"	188'5"	131'11"	94'2"	65'11"	47'1"	33'	24'	16'6"	12'
40	344'5"	246'	172'2"	123'	86'1"	61'6"	43'1"	31'3"	21'5"	15'8"
50	538'2"	384'5"	269'1"	192'2"	134'6"	96'1"	67'3"	48'9'	33'6"	24'5"
60	775'	553'7"	387'6"	276'10"	193'8"	138'5"	96'11"	70'6"	48'5"	35'2"
65	909'6"	649'8"	454'8"	324'10"	227'5"	162'5"	113'8"	82'8"	56'10"	41'4"
75	1211'	865'	605'6"	432'6"	302'8"	216'2"	151'5"	110'1"	75'8"	55'
85	1555'	1111'	777'8"	555'6"	388'10"	277'8"	194'5"	141'5"	97'2"	70'8"
100	2153'	1538'	1076'	768'11"	538'2"	384'5"	269'1"	195'8"	134'6"	97'11"
125	3364'	2403'	1682'	1201'	840'11"	600'8"	420'5"	305'8"	210'2"	152'11"
135	3923'	2802'	1962'	1401'	980'11"	700'8"	490'4"	356'7"	245'2"	178'4"
150	4844'	3460'	2422'	1730'	1211'	865'	605'6"	440'4"	302'8"	220'2"
180	6975'	4982'	3487'	2491'	1744'	1246'	871'11"	634'1"	435'11"	317'

Table E.7 35mm Hyperfocal Distances—Circle of Confusion = .001″

	f-stop									
FOCAL LENGTH	1	1.4	2	2.8	4	5.6	8	11	16	22
5.9	4'6"	3'2"	2'2"	1'7"	1'1"	10"	7"	5"	4"	2"
8	8'4"	5'11"	4'1"	2'11"	2'1"	1'6"	1'	10"	6"	5"
9.5	11'8"	8'4"	5'10"	4'2"	2'11"	2'1"	1'6"	1'1"	9"	6"
10	12'11"	9'2"	6'6"	4'7"	3'2"	2'4"	1'6"	1'2"	10"	7"
12	18'7"	13'4"	9'4"	6'7"	4'7"	3'4"	2'4"	1'8"	1'2"	10"
14	25'4"	18'1"	12'8"	9'	6'4"	4'6"	3'2"	2'4"	1'7"	1'2"
16	33'1"	23'7"	16'6"	11'10"	8'4"	5'11"	4'1"	3'	2'1"	1'6"
17	37'4"	26'8"	18'8"	13'4"	9'4"	6'8"	4'8"	3'5"	2'4"	1'8"
18	41'10"	29'11"	20'11"	14'11"	10'6"	7'6"	5'2"	3'10"	2'7"	1'11"
20	51'8"	36'11"	25'10"	18'6"	12'11"	9'2"	6'6"	4'8"	3'2"	2'4"
21	57'	40'8"	28'6"	20'4"	14'2"	10'2"	7'1"	5'2"	3'7"	2'7
24	74'5"	53'1"	37'2"	26'7"	18'7"	13'4"	9'4"	6'10"	4'7"	3'5"
25	80'8"	57'8"	40'5"	28'10"	20'2"	14'5"	10'1"	7'4"	5'	3'8"
27	94'2"	67'4"	47'1"	33'7"	23'6"	16'10"	11'10"	8'7"	5'11"	4'4"
28	101'4"	72'4"	50'7"	36'2"	20'10"	18'1"	12'8"	9'2"	6'4"	4'7"
29	108'8"	77'7"	54'4"	38'10"	27'2"	19'5"	13'7"	9'11"	6'10"	4'11"
32	132'4"	94'6"	66'1"	47'2"	33'1"	23'7"	16'6"	12'	8'4"	6'
35	158'2"	113'	79'1"	56'6"	39'7"	28'4"	19'10"	14'5"	9'11"	7'2"
40	206'8"	147'7"	103'4"	73'10"	51'8"	36'11"	25'10"	18'10"	12'11"	9'5"
50	322'11"	230'8"	161'6"	115'4"	80'8"	57'8"	40'5"	29'4"	20'2"	14'7"
60	465'	332'1"	232'6"	166'1"	116'2"	83'	58'1"	42'4"	29'1"	21'1'
65	545'8"	389'10"	272'11"	194'11"	136'5"	97'6"	68'2"	49'7"	34'1"	24'10"
75	726'7"	518'11"	363'4"	259'6"	181'7"	129'8"	90'10"	66'1"	45'5"	33'
85	933'2"	666'7"	466'7"	333'4"	233'4"	166'7"	116'8"	84'10"	58'4"	42'5"
100	1292'	922'7"	645'10"	461'4"	322'11"	230'8"	161'6"	117'5"	80'8"	58'8"
125	2018'	1442'	1009'	720'10"	504'7"	360'5"	252'4"	183'6"	126'1"	91'8"
135	2354'	1682'	1177'	840'8"	588'6"	420'5"	294'4"	214'	147'1"	107'
150	2906'	2076'	1453'	1038'	726'7"	519'	363'4"	264'4"	181'7"	132'1"
180	4185'	2989'	2092'	1495'	1046'	747'4"	523'1"	380'6"	261'7"	190'2"

FEET PER SECOND AND FEET PER MINUTE

You should remember that for 16mm format at 24 f.p.s., the film travels through the camera at the rate of 36 feet per minute, for 35mm 3-perf format at 24 f.p.s., the film travels through the camera at the rate of 67.5 feet per minute, and for 35mm 4-perf format at 24 f.p.s., the film travels through the camera at the rate of 90 feet per minute. Since you will not always be filming at sync speed, Table E.8 lists feet per second and feet per minute for various frames per second for each format. For example, when shooting 16mm film at 18 f.p.s., the film travels through the camera at the rate of 27 feet per minute.

RUNNING TIME TO FILM LENGTH AND FILM LENGTH TO RUNNING TIME

You will often be in a situation where you need to determine if you have enough film to complete a certain shot. Tables E.9–E.11 lists the approximate running times for full rolls of film at various frames per

Table E.8 Feet per Second and Feet per Minute

	16mm		35mm 3-perf		35mm 4-perf	
FPS	**FT / SEC**	**FT / MIN**	**FT / SEC**	**FT / MIN**	**FT / SEC**	**FT / MIN**
6 fps	.15	9	.28	16.9	.375	22.5
12 fps	.3	18	.56	33.8	.75	45
18 fps	.45	27	.85	50.7	1.125	67.5
24 fps	.6	36	1.13	67.5	1.5	90
30 fps	.75	45	1.41	84.5	1.875	112.5
36 fps	.9	54	1.69	101.4	2.25	135
48 fps	1.2	72	2.25	135.2	3	180
60 fps	1.5	90	2.82	169	3.75	225
72 fps	1.8	108	3.38	202.8	4.5	270
96 fps	2.4	144	4.5	270.4	6	360
120 fps	3	180	5.63	338	7.5	450

Table E.9 Running Times for 16mm

16mm	100 feet	200 feet	400 feet	1200 feet
FPS	TIME	TIME	TIME	TIME
6 fps	11 min 7 sec	22 min 13 sec	44 min 26 sec	133 min 20 sec
12 fps	5 min 33 sec	11 min 7 sec	22 min 13 sec	66 min 40 sec
18 fps	3 min 42 sec	7 min 24 sec	14 min 49 sec	44 min 26 sec
24 fps	2 min 46 sec	5 min 33 sec	11 min 7 sec	33 min 20 sec
30 fps	2 min 13 sec	4 min 26 sec	8 min 53 sec	26 min 40 sec
36 fps	1 min 51 sec	3 min 42 sec	7 min 25 sec	22 min 13 sec
48 fps	1 min 23 sec	2 min 46 sec	5 min 33 sec	16 min 40 sec
60 fps	1 min 7 sec	2 min 13 sec	4 min 26 sec	13 min 20 sec
72 fps	55 sec	1 min 51 sec	3 min 42 sec	11 min 7 sec
96 fps	41 sec	1 min 23 sec	2 min 46 sec	8 min 20 sec
120 fps	33 sec	1 min 7 sec	2 min 13 sec	6 min 40 sec

Table E.10 Running Times for 35mm 3-Perf Format

35mm 3-perf	100 feet	200 feet	400 feet	1000 feet	2000 feet
FPS	TIME	TIME	TIME	TIME	TIME
6 fps	5 min 55 sec	11 min 50 sec	23 min 40 sec	59 min 10 sec	118 min 18 sec
12 fps	2 min 58 sec	5 min 55 sec	11 min 50 sec	29 min 35 sec	59 min 10 sec
18 fps	1 min 58 sec	3 min 56 sec	7 min 53 sec	19 min 43 sec	39 min 27 sec
24 fps	1 min 29 sec	2 min 58 sec	5 min 56 sec	14 min 49 sec	29 min 38 sec
30 fps	1 min 11 sec	2 min 22 sec	4 min 44 sec	11 min 50 sec	23 min 40 sec
36 fps	59 sec	1 min 58 sec	3 min 56 sec	9 min 52 sec	19 min 43 sec
48 fps	44 sec	1 min 29 sec	2 min 58 sec	7 min 24 sec	14 min 47 sec
60 fps	35 sec	1 min 11 sec	2 min 22 sec	5 min 55 sec	11 min 50 sec
72 fps	29 sec	59 sec	1 min 58 sec	4 min 56 sec	9 min 52 sec
96 fps	22 sec	44 sec	1 min 29 sec	3 min 42 sec	7 min 24 sec
120 fps	18 sec	35 sec	1 min 11 sec	2 min 58 sec	5 min 55 sec

Table E.11 Running Times for 35mm 4-Perf Format

35mm 4-perf	100 feet	200 feet	400 feet	1000 feet	2000 feet
FPS	TIME	TIME	TIME	TIME	TIME
6 fps	4 min 26 sec	8 min 53 sec	17 min 47 sec	44 min 26 sec	88 min 53 sec
12 fps	2 min 13 sec	4 min 26 sec	8 min 53 sec	22 min 13 sec	44 min 26 sec
18 fps	1 min 29 sec	2 min 58 sec	5 min 56 sec	14 min 49 sec	29 min 37 sec
24 fps	1 min 7 sec	2 min 13 sec	4 min 26 sec	11 min 7 sec	22 min 13 sec
30 fps	53 sec	1 min 46 sec	3 min 33 sec	8 min 53 sec	17 min 46 sec
36 fps	44 sec	1 min 29 sec	2 min 58 sec	7 min 24 sec	14 min 48 sec
48 fps	33 sec	1 min 7 sec	2 min 13 sec	5 min 33 sec	11 min 7 sec
60 fps	26 sec	53 sec	1 min 46 sec	4 min 26 sec	8 min 53 sec
72 fps	22 sec	44 sec	1 min 29 sec	3 min 42 sec	7 min 24 sec
96 fps	16 sec	33 sec	1 min 7 sec	2 min 46 sec	5 min 33 sec
120 fps	13 sec	26 sec	53 sec	2 min 13 sec	4 min 26 sec

second. All amounts are rounded to the nearest minute and second. For example, from Table E.10, when shooting 35mm, 3-perf format and using a 400 foot roll, at a speed of 36 f.p.s., the roll will last approximately 3 minutes and 56 seconds

Tables E.12–E.14 show the amount of film used for different times at various frames per seconds. All lengths are rounded to the nearest foot and inch. For example, from Table E.14, when shooting the 35mm, 4-perf format, at 24 f.p.s., a shot that lasts 14 seconds is approximately 21 feet in length.

Tables E.15–E.17 show the amount of time for different film lengths at various frames per second. All amounts are rounded to the nearest minute and second. For example, from Table E.15, when shooting 16mm, at 48 f.p.s., 60 feet of film will last approximately 50 seconds.

Table E.12 Running Time to Film Length—16mm

	frames per second										
seconds	**6**	**12**	**18**	**24**	**30**	**36**	**48**	**60**	**72**	**96**	**120**
1	2"	4"	6"	7"	9"	11"	1'2"	1'6"	1'10"	2'5"	3'
2	4"	7"	11"	1'2"	1'6"	1'10"	2'5"	3'	3'7"	4'10"	6'
3	6"	11"	1'5"	1'10"	2'3"	2'8"	3'7"	4'6"	5'5"	7'2"	9'
4	7"	1'2'	1'10"	2'5"	3'	3'7"	4'10"	6'	7'2"	9'7"	12'
5	9"	1'6"	2'3"	3'	3'9"	4'6"	6'	7'6"	9'	12'	15'
6	11"	1'10"	2'8"	3'7"	4'6"	5'5"	7'2"	9'	10'10"	14'5"	18'
7	1'1"	2'1"	3'2"	4'2"	5'3"	6'4"	8'5"	10'6"	12'7"	16'10"	21'
8	1'2"	2'5"	3'7"	4'10"	6'	7'2"	9'7"	12'	14'5"	19'2"	24'
9	1'4"	2'8"	4'1"	5'5"	6'9"	8'1"	10'10"	13'6"	16'2"	21'7"	27'
10	1'6"	3'	4'6"	6'	7'6"	9'	12'	15'	18'	24'	30'
11	1.8"	3'4"	5'	6'7"	8'3"	9'11"	13'2"	16'6"	19'10"	26'5"	33'
12	1'10"	3'7"	5'5"	7'2"	9'	10'10"	14'5"	18'	21'7"	28'10"	36'
13	2'	3'11"	5'10"	7'10"	9'9"	11'8"	15'7"	19'6"	23'4"	31'2"	39'
14	2'1"	4'2"	6'4"	8'5"	10'6"	12'7"	16'10"	21'	25'2"	33'7"	42'
15	2'3"	4'6"	6'9"	9'	11'3"	13'6"	18'	22'6"	27'	35'	45'
16	2'5"	4'10"	7'2"	9'7"	12'	14'5"	19'2"	24'	28'10"	38'5"	48'
17	2'7"	5'1"	7'8"	10'2"	12'9"	15'4"	20'5"	25'6"	30'7"	40'10"	51'
18	2'8"	5'5"	8'1"	10'10"	13'6"	16'2"	21'6"	27'	32'5"	43'2"	54'
19	2'11"	5'8"	8'7"	11'5"	14'3"	17'1"	22'10"	28'6"	34'2'	45'7"	57'
20	3'	6'	9'	12'	15'	18'	24'	30'	36'	48'	60'
21	3'2"	6'4"	9'6"	12'7"	15'9"	18'11"	25'2"	31'6"	37'8"	50'5"	63'
22	3'4"	6'4"	9'6"	12'7"	15'9"	18'11"	25'2"	31'6"	37'10"	50'5"	63'
23	3'6"	6'11"	10'4"	13'10"	17'3"	20'8"	27'7"	34'6"	41'5"	55'2"	69'
24	3'7"	7'2"	10'10"	14'5"	18'	21'7"	28'10"	36'	43'2"	57'7"	72'
25	3'9"	7'6"	11'3"	15'	18'9"	22'6"	30'	37'6"	45'	60'	75'
26	3'11"	7'10"	11'8"	15'7"	19'6"	23'5"	31'2"	39'	46'7"	64'10"	81'
27	4'1"	8'1"	12'2"	16'2"	20'3"	24'4"	32'5"	40'6"	48'7"	64'10"	81'
28	4'2"	8'5"	12'7"	16'10"	21'	25'2"	33'7"	42'	50'5"	67'2"	84'
29	4'4"	8'8"	13'1"	17'5"	21'9"	26'1"	34'10"	43'6"	52'2"	69'7"	87'
30	4'6"	9'	13'6"	18'	22'6"	27'	36'	45'	54'	72'	90'

Table E.12 *(continued)*

	frames per second										
seconds	6	12	18	24	30	36	48	60	72	96	120
31	4'8"	9'4"	13'11"	18'7"	23'3"	27'11"	37'2"	46'6"	55'10"	74'5"	93'
32	4'10"	9'7"	14'5"	19'2"	24'	28'10"	38'5"	48'	57'7"	76'10"	96'
33	4'11"	9'11"	14'11"	19'10"	24'9"	29'8"	39'7"	49'6"	59'5"	79'2"	99'
34	5'1"	10'2"	15'4"	20'5"	25'6"	30'7"	40'10"	51'	61'2"	81'7"	102'
35	5'3"	10'6"	15'9"	21'	26'3"	31'6"	42'	52'6"	63'	84'	105'
36	5'5"	10'10"	16'2"	21'7"	27'	32'5"	43'2"	54'	64'10"	86'5"	108'
37	5'7"	11'1"	16'7"	22'2"	27'9"	33'4"	44'5"	55'6"	66'7"	88'10"	111'
38	5'8"	11'5"	17'1"	22'10"	28'6"	34'2"	45'7"	57'	68'5"	91'2"	114'
39	5'11"	11'8"	17'7"	23'5"	29'3"	35'1"	46'10"	58'6"	70'2"	93'7"	117'
40	6'	12'	18'	24'	30'	36'	48'	60'	72'	96'	120'
41	6'2"	12'4"	18'6"	24'7"	30'9"	36'11"	49'2"	61'6"	73'10"	98'5"	123'
42	6'4"	12'7"	18'11"	25'2"	31'6"	37'10"	50'5"	63'	75'6"	100'10"	126'
43	6'6"	12'11"	19'4"	25'10"	32'3"	38'8"	51'7"	64'6"	77'5"	103'2"	129'
44	6'7"	13'2"	19'10"	26'5"	33'	39'7"	52'10"	66'	79'2"	105'7"	132'
45	6'9"	13'6"	20'3"	27'	33'9"	40'6"	54'	67'6"	81'	108'	135'
46	6'11"	13'10"	20'8"	27'7"	34'6"	41'5"	55'2"	69'	82'10"	110'5"	138'
47	7'1"	14'1"	21'2"	28'2"	35'3"	42'4"	56'5"	70'6"	84'7"	112'10"	141'
48	7'2"	14'5"	21'7"	28'10"	36'	43'2"	57'6"	72'	86'5"	115'2"	144'
49	7'4"	14'8"	22'1"	29'5"	36'9"	44'1"	58'10"	73'6"	88'2"	117'7"	147'
50	7'6"	15'	22'6"	30'	37'6"	45'	60'	75'	90'	120'	150'
51	7'8"	15'4"	22'11"	30'7"	38'3"	45'11"	61'2"	76'6"	91'10"	122'5"	153'
52	7'10"	15'7"	23'5"	31'2"	39'	46'10"	62'5"	78'	93'7"	124'10"	156'
53	7'11"	15'11"	23'11"	31'10"	39'9"	47'8"	63'7"	79'6"	95'5"	127'2"	159'
54	8'1"	16'2"	24'4"	32'5"	40'6"	48'7"	64'10"	81'	97'2"	129'7"	162'
55	8'3"	16'6"	24'9"	33'	41'3"	49'6"	66'	82'6"	99'	132'	165'
56	8'5"	16'10"	25'2"	33'7"	42'	50'5"	67'2"	84'	100'10"	134'5"	168'
57	8'7"	17'1"	25'8"	34'2"	42'9"	51'4"	68'5"	85'6"	102'7"	136'10"	171'
58	8'8"	17'5"	26'1"	34'10"	43'6"	52'2"	69'7"	87'	104'5"	139'2"	174'
59	8'11"	17'8"	26'7"	35'5"	44'3"	53'1"	70'10"	88'6"	106'2"	141'7"	177'
60	9'	18'	27'	36'	45'	54'	72'	90'	108'	144'	180'

Table E.13 Running Time to Film Length—35mm, 3 Perf

	frames per second										
seconds	**6**	**12**	**18**	**24**	**30**	**36**	**48**	**60**	**72**	**96**	**120**
1	3"	7"	10"	1'2"	1'5"	1'8"	2'3"	2'10"	3'5"	4'6"	5'7"
2	7"	1'2"	1'8"	2'3"	2'10"	3'5"	4'6"	5'7"	6'9"	9'	11'3"
3	10"	1'8"	2'7"	3'5"	4'3"	5'1"	6'9"	8'6"	10'2"	13'6"	16'11"
4	1'1"	2'2"	3'5"	4'6"	5'7"	6'9"	9'	11'3"	13'6"	18'	22'6"
5	1'5"	2'10"	4'4"	5'7"	7'1"	8'6"	11'3"	14'1"	16'11"	22'6"	28'2"
6	1'8"	3'5"	5'1"	6'10"	8'6"	10'2"	13'6"	16'11"	20'3"	27'	33'10"
7	2'	3'11"	6'	7'11"	9'11"	11'10"	15'9"	19'9"	23'8"	31'6"	39'5"
8	2'2"	4'6"	6'10"	9'	11'3"	13'6"	18'	22'7"	27'1"	36'	45'1"
9	2'6"	5'	7'8"	10'1"	12'8"	15'3"	20'3"	25'5"	30'5"	40'6"	50'8"
10	2'10"	5'7"	8'6"	11'4"	14'1"	16'11"	22'6"	28'2"	33'10"	45'	56'4"
11	3'1"	6'2"	9'5"	12'5"	15'6"	18'7"	24'9"	31'1"	37'2"	49'6"	61'11"
12	3'5"	6'8"	10'2"	13'6"	16'11"	20'3"	27'	33'10"	40'7"	54'	67'7"
13	3'7"	7'4"	11'1"	14'7"	18'4"	21'11"	29'3"	36'7"	43'11"	58'6"	73'2"
14	3'11"	7'10"	11'11"	15'10"	19'9"	23'7"	31'6"	39'6"	47'4"	63'	78'10"
15	4'2"	8'5"	12'10"	16'11"	21'2"	25'4"	33'9"	42'4"	50'8"	67'6"	84'6"
16	4'6"	9'	13'7"	18'	22'6"	27'1"	36'	45'2"	54'1"	72'	90'1"
17	4'10"	9'6"	14'6"	19'1"	23'11"	28'9"	38'3"	47'11"	57'6"	76'6"	95'8"
18	5'	10'1"	15'4"	20'4'	25'4"	30'5"	40'6"	50'9"	60'10"	81'	101'4"
19	5'4"	10'7"	16'2"	21'5"	26'9"	32'11"	42'9"	53'7"	64'2"	85'6"	106'11"
20	5'7"	11'2"	17'	22'5"	28'2"	33'10"	45'	56'5"	67'7"	90'	112'7"
21	5'11"	11'10"	17'11"	23'7"	29'7"	35'6"	47'3"	59'2"	71'	94'6"	118'3"
22	6'2"	12'4"	18'8"	24'10"	31'1"	37'2"	49'6"	62'1"	74'4"	99'	123'11"
23	6'5"	12'11"	19'7"	25'11"	32'5"	38'11"	51'9"	64'11"	77'9"	103'6"	129'6"
24	6'8"	13'5"	20'5"	27'	33'10"	40'7"	54'	67'7"	81'2"	108'	135'2"
25	7'	14'	21'4"	28'1"	35'3"	42'3"	56'3"	70'6"	84'6"	112'6"	140'9"
26	7'4"	14'7"	22'1"	29'4"	36'8"	43'11"	58'6"	73'4"	87'11"	117'	146'5"
27	7'7"	15'1"	23'	30'5"	38'1"	45'7"	60'9"	76'2"	91'3"	121'6"	152'
28	7'10"	15'8"	23'10"	31'6"	39'6"	47'4"	63'	78'11"	94'7"	130'6"	163'3"
29	8'1"	16'2"	24'8"	32'7"	40'11"	49'	65'3"	81'9"	98'	130'6"	163'3"
30	8'5"	16'10"	26'6"	33'9"	42'4"	50'8"	67'6"	84'7"	101'5"	135'	168'11"

Table E.13 *(continued)*

	frames per second										
seconds	**6**	**12**	**18**	**24**	**30**	**36**	**48**	**60**	**72**	**96**	**120**
31	8'8"	17'5"	26'5"	34'11"	43'8"	52'5"	69'9"	87'5"	104'10"	139'6"	174'6"
32	9'	17'11"	27'2"	36'	45'1"	54'1"	72'	90'3"	108'2"	144'	180'2"
33	9'2"	18'6"	28'1"	37'1"	46'6"	55'8"	74'3"	93'1'	111'6"	148'6"	185'10"
34	9'6"	19'	28'11"	38'4"	47'11"	57'6"	76'6"	95'11"	114'11"	153'	191'5"
35	9'10"	19'7"	29'10"	39'5"	49'4"	59'2"	78'9"	98'8"	118'4"	157'6"	197'1"
36	10'1"	20'2"	30'7"	40'6"	50'9"	60'9"	81'	101'6"	121'7"	162'	202'7"
37	10'5"	20'8"	31'6"	41'7"	52'2"	62'6"	83'3"	104'4"	125'1"	166'6"	208'4"
38	10'7"	21'4"	32'4"	42'10"	53'7"	64'3"	85'6"	107'2"	128'6"	171'	213'11"
39	10'11"	21'11"	33'2"	43'11"	55'	65'11"	87'9"	110'	131'10"	175'6"	219'7"
40	11'2"	22'5"	34'	45'	56'5"	67'7"	90'	112'10"	135'2"	180'	225'2"
41	11'6"	23'	34'11"	46'1"	57'10"	69'4"	92'3"	115'7"	138'7"	184'6"	230'10"
42	11'10"	23'6"	35'8"	47'4"	59'3"	71'	94'6"	118'6"	142'	189'	236'6"
43	12'	24'1"	36'7"	48'5"	60'7"	72'7"	96'9"	121'3"	145'4"	193'6"	242'1"
44	12'4"	24'7"	37'5"	49'6"	62'	74'4"	99'	124'1"	148'8"	198'	247'9"
45	12'7"	25'2"	38'4"	50'7"	63'6"	76'1"	101'3"	126'11"	152'1"	202'6"	253'4"
46	12'11"	25'10"	39'1"	51'10"	64'11"	77'9"	103'6"	129'9"	155'6"	207'	259'
47	13'2"	26'4"	40'	52'11"	66'3"	79'5"	105'9"	132'6"	158'11"	211'6"	264'7"
48	13'5"	26'11"	40'10"	54'	67'8"	81'2"	108'	135'4"	162'3"	216'	270'3"
49	13'8"	27'5"	41'8"	55'1"	69'1"	82'10"	110'3"	138'2"	165'7"	220'6"	275'11"
50	14'	28'	42'6"	56'4"	70'6"	84'6"	112'6"	141'	169'	225'	281'6"
51	14'4"	28'7"	43'5"	57'5"	71'11"	86'2"	114'9"	143'10"	172'5"	229'6"	287'2"
52	14'7"	29'1"	44'2"	58'6"	73'4"	87'11"	117'	146'7"	175'9"	234'	292'9"
53	14'10"	29'8"	45'1"	59'7"	74'9"	89'7"	119'3"	149'6"	179'2"	238'6"	298'5"
54	15'1"	30'2"	45'11"	60'10"	76'2"	91'3"	121'6"	152'3"	182'6"	243'	304'
55	15'5"	30'10"	46'10"	61'11"	77'6"	93'	123'9"	155'1"	185'11"	247'6"	309'8"
56	15'8"	31'5"	47'7"	63'	79'	94'7"	126'	158'	189'3"	252'	315'3"
57	16'	31'11"	48'6"	64'1"	80'4"	96'4"	128'3"	160'9"	192'8"	256'6"	320'11"
58	16'2"	32'6"	49'4"	65'4"	81'9"	98'	130'6"	163'7"	196'	261'	326'6"
59	16'6"	33'	50'2"	66'5"	83'2"	99'8"	132'9"	166'5"	199'5"	265'6"	332'2"
60	16'10"	33'7"	51'	67'6"	84'7"	101'5"	135'	169'2"	202'10"	270'	337'10"

Table E.14 Running Time to Film Length—35mm, 4-Perf

	frames per second										
seconds	**6**	**12**	**18**	**24**	**30**	**36**	**48**	**60**	**72**	**96**	**120**
1	5"	9"	1'1"	1'6"	1'11"	2'3"	3'	3'9"	4'6"	6'	7'6"
2	9"	1'6"	2'3"	3'	3'9"	4'6"	6'	7'6"	9'	12'	15'
3	1'2"	2'3"	3'5"	4'6"	5'8"	6'9"	9'	11'3"	13'6"	18'	22'6"
4	1'6"	3'	4'6"	6'	7'6"	9'	12'	15'	18'	24'	30'
5	1'11"	3'9"	5'8"	7'6"	9'5"	11'3"	15'	18'9"	22'6"	30'	37'6"
6	2'3"	4'6"	6'9"	9'	11'3"	13'6"	18'	22'6"	27'	36'	45'
7	2'8"	5'3"	7'11"	10'6"	13'2"	15'9"	21'	26'3"	31'6"	42'	52'6"
8	3'	6'	9'	12'	15'	18'	24'	30'	36'	48'	60'
9	3'5"	6'9"	10'2"	13'6"	16'11"	20'3"	27'	33'9"	40'6"	54'	67'6"
10	3'9"	7'6"	11'3"	15'	18'9"	22'6"	30'	37'6"	45'	60'	75'
11	4'2"	8'3"	12'5"	16'6"	20'7"	24'9"	33'	41'3"	49'6"	66'	82'6"
12	4'6"	9'	13'6"	18'	22'6"	27'	36'	45'	54'	72'	90'
13	4'11"	9'9"	14'7"	19'6"	24'5"	29'3"	39'	48'9"	58'6"	78'	97'6"
14	5'3"	10'6"	15'9"	21'	26'3"	31'6"	42'	52'6"	63'	84'	105'
15	5'8"	11'3"	16'11"	22'6"	28'2"	33'9"	45'	56'3"	67'6"	90'	112'6"
16	6'	12'	18'	24'	30'	36'	48'	60'	72'	96'	120'
17	6'5"	12'9"	19'2"	25'6"	31'11"	38'3"	51'	63'9"	76'6"	102'	127'6"
18	6'9"	13'6"	20'3"	27'	33'9"	40'6"	54'	67'6"	81'	108'	135'
19	7'2"	14'3"	21'5"	28'6"	35'7"	42'9"	57'	71'3"	85'6"	114'	142'6"
20	7'6"	15'	22'6"	30'	37'6"	45'	60'	75'	90'	120'	150'
21	7'11"	15'9"	23'7"	31'6"	39'5"	47'3"	63'	78'9"	94'6"	126'	157'6"
22	8'3"	16'6"	24'9"	33'	41'3"	49'6"	66'	82'6"	99'	132'	165'
23	8'8"	17'3"	25'11"	34'6"	43'2"	51'9"	69'	86'3"	103'6"	138'	172'6"
24	9'	18'	27'	36'	45'	54'	72'	90'	108'	144'	180'
25	9'5"	18'9"	28'2"	37'6"	46'11"	56'3"	75'	93'9"	112'6"	150'	187'6"
26	9'9"	19'6"	29'3"	39'	48'9"	58'6"	78'	97'6"	117'	156'	195'
27	10'2"	20'3"	30'5"	40'6"	50'7"	60'9"	81'	101'4"	121'6"	162'	202'6"
28	10'6"	21'	31'6"	42'	52'6"	63'	84'	105'	126'	168'	210'
29	10'11"	21'9"	32'7"	43'6"	54'5"	65'3"	87'	108'10"	130'6"	174'	217'6"
30	11'3"	22'6"	33'9"	45'	56'3"	67'6"	90'	112'6"	135'	180'	225'

Table E.14 (*continued*)

	frames per second										
seconds	**6**	**12**	**18**	**24**	**30**	**36**	**48**	**60**	**72**	**96**	**120**
31	11'8"	23'3"	34'11"	46'6"	58'2"	69'9"	93'	116'4"	139'6"	186'	232'6"
32	12'	24'	36'	48'	60'	72'	96'	120'	144'	192'	240'
33	12'5"	24'9"	37'2"	49'6"	61'11"	74'3"	99'	123'10"	148'6"	198'	247'6"
34	12'9"	25'6"	38'3"	51'	63'9"	76'6"	102'	127'5"	153'	204'	255'
35	13'2"	26'3"	39'5"	52'6"	65'8"	78'9"	105'	131'4"	157'6"	210'	262'6"
36	13'6"	27'	40'6"	54'	67'6"	81'	108'	135'	162'	216'	270'
37	13'11"	27'9"	41'8"	55'6"	69'5"	83'3"	111'	138'10"	166'6"	222'	277'6"
38	14'3"	28'6"	42'9"	57'	71'3"	85'6"	114'	142'6"	171'	228'	285'
39	14'8"	29'3"	43'11"	58'6"	73'2"	87'9"	117'	146'4"	175'6"	234'	292'6"
40	15'	30'	45'	60'	75'	90'	120'	150'	180'	240'	300'
41	15'5"	30'9"	46'2"	61'6"	76'11"	92'3"	123'	153'10"	184'6"	246'	307'6"
42	15'9"	31'6"	47'3"	63'	78'9"	94'6"	126'	157'6"	189'	252'	315'
43	16'2"	32'3"	48'5"	64'6"	80'8"	96'9"	129'	161'4"	193'6"	258'	322'6"
44	16'6"	33'	49'6"	66'	82'6"	99'	132'	165'	198'	264'	330'
45	16'11"	33'9"	50'8"	67'6"	84'5"	101'4"	135'	168'10"	202'6"	270'	337'6"
46	17'3"	34'6"	51'9"	69'	86'3"	103'6"	138'	172'6"	207'	276'	345'
47	17'8"	35'3"	52'11"	70'6"	88'2"	105'10"	141'	176'4"	211'6"	282'	352'5"
48	18'	36'	54'	72'	90'	108'	144'	180'	216'	288'	360'
49	18'5"	36'9"	55'2"	73'6"	91'11"	110'4"	147'	183'10"	225'	300'	375'
50	18'9"	37'6"	56'3"	75'	93'9"	112'6"	150'	187'6"	225'	300'	375'
51	19'2"	38'3"	57'5"	76'6"	95'8"	114'10"	153'	191'4"	229'6"	306'	382'6"
52	19'6"	39'	58'6"	78'	97'6"	117'	156'	195'	234'	312'	390'
53	19'11"	39'9"	59'8"	79'6"	99'5"	119'4"	159'	198'10"	238'6"	318'	397'6"
54	20'3"	40'6"	60'9"	81'	101'4"	121'6"	162'	202'6"	243'	324'	405'
55	20'8"	41'3"	61'11"	82'6"	103'1"	123'10"	165'	206'3"	247'6"	330'	412'6"
56	21'	42'	63'	84'	105'	126'	168'	210'	252'	336'	420'
57	21'5"	42'9"	64'2"	85'6"	106'11"	128'4"	171'	213'10"	256'6"	342'	427'6"
58	21'9"	43'6"	65'3"	87'	108'10"	130'6"	174'	217'6"	261'	348'	435'
59	22'2"	44'3"	66'5"	88'6"	110'7"	132'10"	177'	221'4"	265'6"	354'	442'6"
60	22'6"	45'	67'6"	90'	112'6"	135'	180'	225'	270'	360'	450'

Table E.15 Film Length to Running Time—16mm

	fps										
feet	**6**	**12**	**18**	**24**	**30**	**36**	**48**	**60**	**72**	**96**	**120**
1	7 sec	3 sec	2 sec	2 sec	1 sec	1 sec	1 sec	.7 sec	.6 sec	.4 sec	.3 sec
2	13 sec	7 sec	4 sec	3 sec	3 sec	2 sec	2 sec	1 sec	1 sec	1 sec	.7 sec
3	20 sec	10 sec	7 sec	5 sec	4 sec	3 sec	3 sec	2 sec	2 sec	1 sec	1 sec
4	27 sec	13 sec	9 sec	7 sec	5 sec	4 sec	3 sec	3 sec	2 sec	2 sec	1 sec
5	33 sec	17 sec	11 sec	8 sec	7 sec	6 sec	4 sec	3 sec	3 sec	2 sec	2 sec
6	40 sec	20 sec	13 sec	10 sec	8 sec	7 sec	5 sec	4 sec	3 sec	3 sec	2 sec
7	47 sec	23 sec	16 sec	12 sec	9 sec	8 sec	6 sec	5 sec	4 sec	3 sec	2 sec
8	53 sec	27 sec	18 sec	13 sec	11 sec	9 sec	7 sec	5 sec	4 sec	3 sec	3 sec
9	1 min	30 sec	20 sec	15 sec	12 sec	10 sec	7 sec	6 sec	5 sec	4 sec	3 sec
10	1 min 7 sec	33 sec	22 sec	17 sec	13 sec	11 sec	8 sec	7 sec	6 sec	4 sec	3 sec
20	2 min 13 sec	1 min 7 sec	44 sec	33 sec	27 sec	22 sec	17 sec	13 sec	11 sec	8 sec	7 sec
30	3 min 20 sec	1 min 40 sec	1 min 7 sec	50 sec	40 sec	33 sec	25 sec	20 sec	17 sec	13 sec	10 sec
40	4 min 26 sec	2 min 13 sec	1 min 29 sec	1 min 7 sec	53 sec	44 sec	33 sec	27 sec	22 sec	17 sec	13 sec
50	5 min 33 sec	2 min 47 sec	1 min 51 sec	1 min 23 sec	1 min 6 sec	56 sec	42 sec	34 sec	28 sec	21 sec	17 sec
60	6 min 40 sec	3 min 20 sec	2 min 13 sec	1 min 40 sec	1 min 20 sec	1 min 7 sec	50 sec	40 sec	33 sec	25 sec	20 sec
70	7 min 47 sec	3 min 53 sec	2 min 35 sec	1 min 57 sec	1 min 33 sec	1 min 18 sec	58 sec	47 sec	39 sec	29 sec	23 sec
80	8 min 54 sec	4 min 26 sec	2 min 58 sec	2 min 14 sec	1 min 46 sec	1 min 29 sec	1 min 6 sec	54 sec	44 sec	33 sec	26 sec
90	10 min	5 min	3 min 20 sec	2 min 30 sec	2 min	1 min 40 sec	1 min 15 sec	1 min	50 sec	38 sec	30 sec
100	11 min 7 sec	5 min 33 sec	3 min 42 sec	2 min 47 sec	2 min 13 sec	1 min 51 sec	1 min 23 sec	1 min 7 sec	55 sec	41 sec	33 sec
200	22 min 14 sec	11 min 6 sec	7 min 24 sec	5 min 34 sec	4 min 26 sec	3 min 42 sec	2 min 46 sec	2 min 13 sec	1 min 51 sec	1 min 23 sec	1 min 7 sec
400	44 min 28 sec	22 min 12 sec	14 min 48 sec	11 min 7 sec	8 min 53 sec	7 min 25 sec	5 min 33 sec	4 min 26 sec	3 min 42 sec	2 min 46 sec	2 min 13 sec
1200	133 min 24 sec	66 min 36 sec	44 min 24 sec	33 min 20 sec	26 min 40 sec	22 min 13 sec	16 min 40 sec	13 min 20 sec	11 min 7 sec	8 min 20 sec	6 min 40 sec

Table E.16 Film Length to Running Time—35mm, 3-Perf

	fps										
feet	**6**	**12**	**18**	**24**	**30**	**36**	**48**	**60**	**72**	**96**	**120**
1	4 sec	2 sec	1 sec	.9 sec	.7 sec	.6 sec	.4 sec	.4 sec	.3 sec	.2 sec	.2 sec
2	7 sec	4 sec	2 sec	2 sec	1 sec	1 sec	.9 sec	.7 sec	.6 sec	.4 sec	.4 sec
3	11 sec	5 sec	4 sec	4 sec	2 sec	2 sec	1 sec	1 sec	.9 sec	.7 sec	.5 sec
4	14 sec	7 sec	5 sec	4 sec	3 sec	2 sec	2 sec	1 sec	1 sec	.9 sec	.7 sec
5	18 sec	9 sec	6 sec	4 sec	4 sec	3 sec	2 sec	2 sec	1 sec	1 sec	.9 sec
6	21 sec	11 sec	7 sec	5 sec	4 sec	4 sec	3 sec	2 sec	2 sec	1 sec	1 sec
7	25 sec	12 sec	8 sec	6 sec	5 sec	4 sec	3 sec	3 sec	2 sec	2 sec	1 sec
8	28 sec	14 sec	9 sec	7 sec	6 sec	5 sec	4 sec	3 sec	2 sec	2 sec	1 sec
9	32 sec	16 sec	11 sec	8 sec	6 sec	5 sec	4 sec	3 sec	2 sec	2 sec	2 sec
10	36 sec	18 sec	12 sec	9 sec	7 sec	6 sec	4 sec	4 sec	3 sec	2 sec	2 sec
20	1 min 11 sec	36 sec	24 sec	18 sec	14 sec	12 sec	9 sec	7 sec	6 sec	4 sec	4 sec
30	1 min 47 sec	53 sec	35 sec	26 sec	21 sec	18 sec	13 sec	11 sec	9 sec	7 sec	5 sec
40	2 min 22 sec	1 min 11 sec	47 sec	35 sec	28 sec	24 sec	18 sec	14 sec	12 sec	9 sec	7 sec
50	2 min 57 sec	1 min 29 sec	59 sec	44 sec	36 sec	30 sec	22 sec	18 sec	15 sec	11 sec	9 sec
60	3 min 33 sec	1 min 47 sec	1 min 11 sec	53 sec	43 sec	35 sec	26 sec	22 sec	18 sec	13 sec	11 sec
70	4 min 12 sec	2 min 5 sec	1 min 23 sec	1 min 2 sec	50 sec	41 sec	31 sec	25 sec	21 sec	15 sec	13 sec
80	4 min 44 sec	2 min 22 sec	1 min 34 sec	1 min 10 sec	57 sec	47 sec	35 sec	29 sec	24 sec	18 sec	14 sec
90	5 min 20 sec	2 min 40 sec	1 min 46 sec	1 min 19 sec	1 min 4 sec	53 sec	40 sec	32 sec	27 sec	20 sec	16 sec
100	5 min 55 sec	2 min 58 sec	1 min 58 sec	1 min 29 sec	1 min 11 sec	59 sec	44 sec	35 sec	30 sec	22 sec	18 sec
200	11 min 50 sec	5 min 55 sec	3 min 56 sec	2 min 58 sec	2 min 22 sec	1 min 58 sec	1 min 29 sec	1 min 11 sec	59 sec	44 sec	35 sec
400	23 min 40 sec	11 min 50 sec	7 min 53 sec	5 min 56 sec	4 min 44 sec	3 min 56 sec	2 min 58 sec	2 min 22 sec	1 min 58 sec	1 min 29 sec	1 min 11 sec
1000	59 min 10 sec	29 min 35 sec	19 min 43 sec	14 min 49 sec	11 min 50 sec	9 min 52 sec	7 min 24 sec	5 min 55 sec	4 min 56 sec	3 min 42 sec	2 min 58 sec
2000	118 min 18 sec	59 min 10 sec	39 min 27 sec	29 min 38 sec	23 min 40 sec	19 min 43 sec	14 min 47 sec	11 min 50 sec	9 min 52 sec	7 min 24 sec	5 min 55 sec

Table E.17 Film Length to Running Time—35mm, 4-Perf

	fps										
feet	**6**	**12**	**18**	**24**	**30**	**36**	**48**	**60**	**72**	**96**	**120**
1	3 sec	1 sec	.9 sec	.7 sec	.5 sec	.4 sec	.3 sec	.3 sec	.2 sec	.2 sec	.1 sec
2	5 sec	3 sec	2 sec	1 sec	1 sec	.9 sec	.7 sec	.5 sec	.4 sec	.3 sec	.3 sec
3	8 sec	4 sec	3 sec	2 sec	2 sec	1 sec	1 sec	.8 sec	.7 sec	.5 sec	.4 sec
4	11 sec	5 sec	4 sec	3 sec	2 sec	2 sec	1 sec	1 sec	.9 sec	.7 sec	.5 sec
5	13 sec	7 sec	4 sec	3 sec	3 sec	2 sec	2 sec	1 sec	1 sec	.9 sec	.7 sec
6	16 sec	8 sec	5 sec	4 sec	3 sec	3 sec	2 sec	2 sec	1 sec	1 sec	.8 sec
7	19 sec	9 sec	6 sec	5 sec	4 sec	3 sec	2 sec	2 sec	2 sec	1 sec	.9 sec
8	21 sec	11 sec	7 sec	5 sec	4 sec	4 sec	3 sec	2 sec	2 sec	1 sec	1 sec
9	24 sec	12 sec	8 sec	6 sec	5 sec	4 sec	3 sec	2 sec	2 sec	2 sec	1 sec
10	27 sec	13 sec	9 sec	7 sec	5 sec	4 sec	3 sec	3 sec	2 sec	2 sec	1 sec
20	53 sec	27 sec	18 sec	13 sec	11 sec	9 sec	7 sec	5 sec	4 sec	3 sec	3 sec
30	1 min 20 sec	40 sec	26 sec	20 sec	16 sec	13 sec	10 sec	8 sec	7 sec	5 sec	4 sec
40	1 min 47 sec	53 sec	35 sec	27 sec	21 sec	18 sec	13 sec	11 sec	9 sec	7 sec	5 sec
50	2 min 14 sec	1 min 6 sec	44 sec	34 sec	27 sec	22 sec	17 sec	14 sec	11 sec	9 sec	7 sec
60	2 min 40 sec	1 min 20 sec	53 sec	40 sec	32 sec	26 sec	20 sec	16 sec	13 sec	10 sec	8 sec
70	3 min 7 sec	1 min 33 sec	1 min 2 sec	47 sec	37 sec	31 sec	23 sec	19 sec	15 sec	12 sec	9 sec
80	3 min 34 sec	1 min 46 sec	1 min 10 sec	54 sec	42 sec	35 sec	26 sec	22 sec	18 sec	14 sec	10 sec
90	4 min	2 min	1 min 19 sec	1 min	48 sec	40 sec	30 sec	24 sec	20 sec	15 sec	12 sec
100	4 min 26 sec	2 min 13 sec	1 min 29 sec	1 min 7 sec	53 sec	44 sec	33 sec	26 sec	22 sec	16 sec	13 sec
200	8 min 53 sec	4 min 26 sec	2 min 58 sec	2 min 13 sec	1 min 46 sec	1 min 29 sec	1 min 7 sec	53 sec	44 sec	33 sec	26 sec
400	17 min 47 sec	8 min 53 sec	5 min 56 sec	4 min 26 sec	3 min 33 sec	2 min 58 sec	2 min 13 sec	1 min 46 sec	1 min 29 sec	1 min 7 sec	53 sec
1000	44 min 26 sec	22 min 13 sec	14 min 49 sec	11 min 7 sec	8 min 53 sec	7 min 24 sec	5 min 33 sec	4 min 26 sec	3 min 42 sec	2 min 46 sec	2 min 13 sec
2000	88 min 53 sec	44 min 26 sec	29 min 37 sec	22 min 13 sec	17 min 46 sec	14 min 48 sec	11 min 7 sec	8 min 53 sec	7 min 24 sec	5 min 33 sec	4 min 26 sec

Recommended Reading

Bernstein, Steven. *The Technique of Film Production.* Newton, MA: Focal Press. 1988.

Bloedow, Jerry. *Filmmaking Foundations.* Newton, MA: Focal Press. 1991.

Brown, Blain. *The Filmmaker's Pocket Reference.* Newton, MA: Focal Press. 1994.

Browne, Steven E. *Film-Video Terms and Concepts.* Newton, MA: Focal Press. 1992.

Carlson, Verne, and Sylvia Carlson. *Professional Cameraman's Handbook*, 4th ed. Newton, MA: Focal Press. 1994.

Carlson, Verne, and Sylvia Carlson. *Professional Lighting Handbook*, 2nd ed. Newton, MA: Focal Press. 1991.

Cheshire, David. *The Book of Movie Photography.* New York: Alfred A. Knopf. 1979.

Coe, Brian. *The History of Movie Photography.* New York: Zoetrope. 1982.

Courter, Philip R. *The Filmmakers Craft: 16mm Cinematography.* New York: Van Nostrand Reinhold Co. 1982.

Daley, Ken. *Basic Film Technique.* Newton, MA: Focal Press. 1980.

Dancyger, Ken. *The Technique of Film and Video Editing.* Newton, MA: Focal Press. 1993.

Dmytryk, Edward. *Cinema: Concept & Practice.* Newton, MA: Focal Press. 1988.

Eastman Kodak Co. *Handbook of Kodak Photographic Filters.* Rochester, NY: Eastman Kodak Co. 1990.

Eastman Kodak Co. *Eastman Professional Motion Picture Films.* Rochester, NY: Eastman Kodak Co. 1992.

Fauer, Jon. *The Arri 35 Book.* Blauvelt, NY: Arriflex Corp. 1989.

Fauer, Jon. *The 16SR Book 2nd. ed.* Blauvelt, NY: Arriflex Corp. 1992.

Ferncase, Richard K. *Film and Video Lighting Terms and Concepts.* Newton, MA: Focal Press. 1994.

Fielding, Raymond. *A Technological History of Motion Pictures and Television.* Berkeley, CA: University of California Press. 1967.

Garvey, Helen. *Before You Shoot.* Santa Cruz, CA: Shire Press. 1985.

Happe, L. Bernard. *Basic Motion Picture Technology.* Newton, MA: Focal Press. 1971.

Harrison, H. K. *Mystery of Filters II.* Porterville, CA: Harrison & Harrison. 1981.

Hart, Douglas C. *The Camera Assistant: A Complete Professional Handbook.* Newton, MA: Focal Press. 1995.

Hershey, Fritz. *Optics and Focus for Camera Assistants.* Newton, MA: Focal Press. 1995.

Hirschfeld, Gerald, A.S.C. *Image Control.* Newton, MA: Focal Press. 1993.

Kindem, Gorham. *The Moving Image: Production Principles and Practice.* Glenview, IL: Scott, Foresman. 1987.

Lowell, Ross. *Matters of Light & Depth.* Philadelphia, PA: Broad Street Books. 1992.

Macdonald, Scott. *A Critical Cinema.* Berkeley, CA: University of California Press. 1988.

Malkiewicz, Kris, and Robert E. Rogers. *Cinematography.* New York: Prentice-Hall. 1988.

Malkiewicz, Kris, and Robert E. Rogers. *Film Lighting.* New York: Prentice-Hall. 1986.

Maltin, Leonard. *The Art of the Cinematographer: A Survey and Interview with Five Masters.* New York: Dover. 1978.

Miller, Pat P. *Script Supervising and Film Continuity.* Newton, MA: Focal Press. 1990.

Penney, Edmund F. *The Facts on File Dictionary of Film and Broadcast Terms.* New York: Facts on File, 1991.

Pincus, Edward, and Steven Ascher. *The Filmmaker's Handbook.* New York: New American Library. 1984.

Roberts, Kenneth H., and Win Sharples, Jr. *A Primer for Filmmaking: A Complete Guide to 16 and 35mm Film Production.* New York: Bobbs-Merrill. 1971.

Samuelson, David W. *Motion Picture Camera Data.* Newton, MA: Focal Press. 1979.

Samuelson, David W. *Motion Picture Camera Techniques.* Newton, MA: Focal Press. 1984.

Samuelson, David W. *Motion Picture Camera and Lighting Equipment.* Newton, MA: Focal Press. 1987.

Samuelson, David W. *Panaflex Users' Manual.* Newton, MA: Focal Press. 1990.

Samuelson, David W. *'Hands-On' Manual for Cinematographers.* Newton, MA: Focal Press. 1994.

Singleton, Ralph S. *Filmmaker's Dictionary.* Beverly Hills, CA: Lone Eagle Publishing Co. 1990.

Taub, Eric. *Gaffers, Grips and Best Boys.* New York: St. Martin's Press. 1987.

Ryan, Rod (ed.). *American Cinematographer Manual.* Hollywood, CA: ASC Press. 1993.

Schaefer, Dennis, and Larry Salvato. *Masters of Light: Conversations with Contemporary Cinematographers.* Berkeley, CA: University of California Press. 1985.

Schroeppel, Tom. *The Bare Bones Camera Course for Film and Video* (2d rev. ed.). Tampa, FL: Tom Schroeppel. 1982.

Souto, Mario Raimondo. *The Technique of the Motion Picture Camera.* Newton, MA: Focal Press. 1969.

Underdahl, Douglas. *The 16mm Camera Book.* New York: Media Logic. 1993.

Uva, Michael. *The Grip Book.* Los Angeles, CA: Michael G. Uva. 1988.

Wheeler, Leslie J. *Principles of Cinematography.* Indianola, IN: Fountain Press. 1953.

Wilson, Anton. *Anton Wilson's Cinema Workshop.* Hollywood, CA: ASC Holding Corp. 1983.

Glossary

Aaton Trade name of a brand of professional 16mm and 35mm camera.
A.C. Abbreviation for Assistant Cameraman.
Academy Aperture The image size of a frame of 35mm motion picture film. It is the ratio of the width to height and is written as 1.33:1.
Acetate Base A film base made up of a slow-burning chemical substance that is much more durable than the older nitrate film base, which was highly flammable. Film that is coated onto an acetate base is sometimes referred to as *safety film*.
Aerial Shot Any filmed shot that is usually done from a plane or helicopter.
A.K.S. A slang term used to refer to an assortment of equipment, tools, and accessories. Any case that contains many different pieces of equipment or accessories is called an AKS case.
American Cinematographer's Manual See A.S.C. Manual.
Anamorphic Lens A film lens that produces an image that is squeezed or compressed to fit the film frame. The developed print of the film is projected through a special projector using a similar type lens, which unsqueezes the image and makes it appear normal on the screen. The anamorphic aspect ratio is 2.40:1.
Angle of View The angle covered by the camera lens. It may also be called *field of view*.
Anti-Halation Backing The dark coating on the back of the unexposed film stock. It is there to prevent light from passing through the film, striking the back of the aperture, and then going back through the film, causing a flare or fogging of the film image.
Aperture (Camera) The opening in the film gate or aperture plate that determines the precise area of exposure of the frame of film.

Aperture (Lens) The opening in the lens, formed by an adjustable iris, through which light passes in order to expose the film. The size of this opening is expressed as an f-stop number.
Aperture Plate A metal plate within the camera that contains an opening in front of the film, which determines the size of the frame.
Arriflex Trade name for a brand of professional 16mm and 35mm camera.
ASA Abbreviation for the American Standards Association. See ASA Speed Rating.
ASA Speed Rating A rating assigned to each film stock to indicate the speed of the film, or the sensitivity of the film to light. The higher the number, the more sensitive the film is to light. A film stock with a low ASA rating may be referred to as a *slow film* and one with a high ASA rating may be referred to as *fast film* or *high speed film*. ASA may also be referred to as E.I. or Exposure Index. See Exposure Index.
A.S.C. Abbreviation for American Society of Cinematographers.
A.S.C. Manual A technical manual that is published by the American Society of Cinematographers. It contains useful information needed by Directors of Photography and Camera Assistants during shooting. This includes information on cameras, lighting, filters, depth of field, exposure compensation, film speed tables, etc.
Aspect Ratio The relationship between the width of the frame to the height of the frame. For television and 16mm films, the standard aspect ratio is 1.33:1, for standard theatrical 35mm feature films it is 1.85:1, and for anamorphic 35mm films it is 2.40:1.
Aspheron A 16mm lens attachment that is designed for the 9.5mm and the 12mm Zeiss prime lenses. It is used to increase the angle of view of these wide angle lenses.
Assistant Cameraman (A.C.) A member of the camera crew who works closely with the Director of Photography and Camera Operator during the shooting day. Some of the job responsibilities include: maintaining and setting up the camera, changing lenses, loading film, measuring focus distances, focusing and zooming during the shot, clapping the slate, placing tape marks for actors, keeping camera reports and other paperwork, etc.

B & W Abbreviation for Black and White.
Baby Legs, Baby Tripod, Babies A short tripod used for low angle shots or any shots where the standard size tripod is not appropriate.
Barney A flexible, padded, and insulated cover used to reduce noise coming from the camera or magazine. A heater version is used

to keep the camera and magazine warm in extremely cold shooting situations.

Base The smooth transparent surface on which the film emulsion is attached. In the earlier days of filmmaking, a nitrate base was used, which was highly flammable. Today a safer acetate-type base is used for all film stocks.

Batteries Rechargeable power supply used to power the camera. Belt batteries and onboard batteries are usually used when doing handheld shots. Block batteries are used when working on a dolly or tripod. Most camera batteries are either 12 volts or 24 volts, depending on the camera system you are using.

Battery Belt A belt containing the cells of the battery, which may be worn by the operator or the assistant when doing handheld shots. It may also be used when a block battery is impractical.

Battery Cables Power cables that are used to supply power from the battery to the camera or accessory being used.

Battery Chargers Electrical device used to keep the batteries fully charged when not being used. When charging, the battery is connected to the charger and the charger is plugged into a standard 110 volt electrical outlet. Never connect the battery to the camera while it is being charged.

Black & White (B & W) It refers to any film shot without color. Sometimes a film may be shot in color, and during the developing and printing process the color is taken away to give a black & white image.

Black Bag A small plastic or paper bag that contains the raw film stock when it is inside the film can. After the film has been exposed, it is placed back in the black bag, and then in the film can, and sealed for delivery to the lab. Some Camera Assistants also refer to the changing bag as the black bag.

Black Dot Texture Screen A diffusion filter that looks like a clear piece of glass containing small black dots in a random pattern. They come in a set ranging from number 1, which is the lightest, to number 5, which is the heaviest. It requires an exposure compensation of one stop.

Block Battery A large camera battery that is enclosed in some type of case containing the cells of the battery as well as a built-in charger. Block batteries may come in single blocks or dual blocks.

Camera The basic piece of equipment used to photograph the images. All cameras consist of a lens that projects the image onto the film stock, a shutter to regulate the light striking the film, a viewfinder that enables the Camera Operator to view the image during filming, some type of mechanism to transport the film through the camera, a motor to power the camera, and a lightproof container, called a *magazine*, that holds the film before and after exposure.

Camera Angle The position of the camera in relation to the subject being filmed (high, low, left, right, etc.).

Camera, Handheld A camera that has been set up so that the Camera Operator may hold it on his shoulders during filming. It may be used to film moving shots or point of view shots of an actor walking or moving through the scene. When filming handheld shots, it is customary to use a wide-angle lens on the camera to minimize the shakiness associated with handheld shots.

Camera Jam A malfunction that occurs when the film backs up in the camera and becomes piled up in the camera movement. The film may become caught between the sprocket wheels and the guide rollers. This may be caused by torn perforations or improper threading in the camera or magazine.

Camera Left The area to the left side of camera as seen from the Camera Operator's point of view. As the actor faces the camera, camera left is to the actor's right.

Camera Mount Any type of device that the camera is mounted on for support. It may be mounted on a head and placed on a dolly, tripod, high hat, camera car, etc.

Cameraman See Director of Photography.

Camera Oil A special type of oil used for lubricating the movement in the camera. It is usually supplied by the camera rental house or camera manufacturer.

Camera Operator The member of the camera crew who looks through and operates the camera during filming. He maintains the composition of the shot, as instructed by the Director and Director of Photography, by making smooth pan and tilt moves.

Camera Rental House A company that specializes in the rental and maintenance of motion picture camera equipment.

Camera Report A form that is filled in with the pertinent information for each roll of film shot. It includes all production information, such as company, title, Director's name, and the Cameraman's name. It should also include the following information: magazine number, roll number, amount of film in the magazine, emulsion number, date shot, scene and take numbers, and the footage used for each roll.

Camera Right The area to the right side of camera as seen from the Camera Operator's point of view. As the actor faces the camera, camera right is to the actor's left.

Camera Speed The rate at which the film is transported through the camera during filming. It is expressed in frames per second, abbreviated f.p.s. Normal sync camera speed is 24 f.p.s.
Camera Tape (1″) Cloth tape, usually one inch in width, that is used for making labels on cases, film cans, and magazines, and also for wrapping cans of exposed film, unexposed film, and short ends. It may also be used by the First Camera Assistant for focus marks. The most commonly used colors of camera tape are white and black, but it is also available in red, yellow, orange, blue, and gray.
Camera Truck A large enclosed truck used to transport and store all camera equipment when filming on location. It is usually set up with a workbench, shelves for storage of equipment, and a darkroom for loading and unloading film.
Camera Wedge A small, wooden wedge that may be used to help level the camera when it is placed on uneven surfaces.
Changing Bag A lightproof cloth bag used to load and unload film when a darkroom is not available.
Changing Tent Very similar in design to a changing bag, except that it forms a dome-shaped tent over the working surface, for loading and unloading film.
Cinematographer See Director of Photography.
Cinematography The art and craft of recording images on motion picture film.
Clap Sticks Term used to refer to the wooden sticks attached to the slate that are clapped together at the beginning of a sync sound take. See Slate.
Clapper Board See Slate.
Clapper/Loader A member of the camera crew who is responsible for clapping the slate for the shot, and also for loading and unloading the film in the magazines. This term is used primarily in Britain and Europe. In the United States it is the Second Camera Assistant.
Closing Down the Lens Turning the diaphragm adjustment ring on the lens to a higher f-stop number, which results in a smaller diaphragm opening.
Coaxial Cable See Video Cables.
Coaxial Magazine A magazine that contains two distinct compartments, on opposite sides of the magazines, one for the feed side and another for the take-up side. *Coaxial* refers to the fact that these two distinct compartments share the same axis of rotation.
Collapsible Core A permanent core in the take-up side of the film magazine, for winding the film onto, after it has been exposed. It has

a slot cut into it, which the end of the film slips into, and a small locking lever that pinches the end of the film against the inside of the core, holding it in place.
Color Chart A card or chart containing strips of colors corresponding to the colors of the spectrum that is used by the lab to assist in developing and processing the film.
Color Compensating Filter A filter that is one of the primary or complementary colors used to make very small adjustments in the color temperature. It comes in varying densities and requires an exposure compensation, depending on the color and density of the filter.
Color Grad Filter Filter that is half-color and half-clear. Used when a specific color effect is desired.
Color Temperature A measurement or scale in degrees Kelvin that measures the specific color of a light.
Combination Filter Two different filter types that are combined into one filter. The most commonly used combination filters are the 85 combined with a neutral density (85N3, 85N6, 85N9) or an 85 combined with a polarizer (85POLA).
"Common Marker" What the Second Camera Assistant calls out when slating a shot for two or more cameras by using only one slate. When using only one slate, all cameras point toward the slate at the beginning of the shot.
"Common Slate" See Common Marker.
Compressed Air Canned air used for blowing out the magazine and camera body. Also used to clean dust off lenses and filters.
Conversion Filter A filter used to convert one color temperature to another. The two most common conversion filters are the 85 and the 80A. The 85 converts tungsten-balanced film for use in daylight and the 80A filter converts daylight-balanced film for use in tungsten light. See 85 filter and 80A filter.
Coral Filter Filter that is used to warm up the overall scene and to enhance skin tones. It is also used to make slight adjustments in Kelvin temperature for different times of day. It requires an exposure compensation depending on the density of the filter.
Core Plastic disks around which the raw stock film is wound. They can be either two or three inches in diameter.
Cotton Swabs Long wooden sticks with a small piece of cotton wrapped around one end which can be used to remove excess oil when oiling the camera. Commonly referred to as **Q-Tips**.
Crosshairs A cross shape that is located on the ground glass of the camera's viewing system. The cross position is in the exact cen-

ter of the film frame to assist the Camera Operator in framing the shot.

Crystal Motor The most common type of camera motor for motion picture cameras. A built-in crystal allows the motor to run at precise speeds, especially when filming with sound, without the use of a cable running from the camera to the sound recorder.

Dailies The developed and printed scenes from the previous day's filming that are viewed by the key production personnel each day. The key personnel includes the Director, Director of Photography, Camera Operator, First Camera Assistant, Editor, etc. Also called *rushes.*

Daily Film Inventory A form that is filled in with information relating to how much film is shot each day. It lists all film stocks and roll numbers used for the day, with a breakdown of good and no-good takes, waste footage, and any short ends made and film on hand at the end of each day.

Darkroom A small, lightproof room, usually 4 ft. by 4 ft. in size, on a stage or in a camera truck, which is used for the loading and unloading of film. The unexposed raw stock film is usually stored in the darkroom during production.

Daylight A light source with a color temperature of approximately 5600 degrees Kelvin.

Daylight Spool A special reel, usually made of metal or plastic, containing opaque edges, onto which the raw stock is tightly wound. It allows the film to be loaded into the camera in daylight, so that only the outer layers of the film will become exposed to the light.

Depth of Field The range of distance within which all objects will be in acceptable sharp focus. It is an area in front of and behind the principle point of focus that will also be in acceptable focus. There is more depth of field behind the point of focus than there is in front of it.

Diaphragm The adjustable metal blades within the lens that control the size of the opening through which the light enters the lens. It may also be called an *iris.* The size of the opening is expressed by an f-stop number.

Diffusion Filter Filter that is used to slightly decrease the sharpness of the image. It is very good for smoothing out facial blemishes or wrinkles. It can also be used for dreamlike effects, and may give the appearance that the image is out of focus. Diffusion filters are available in various densities. Most diffusion filters require no exposure compensation.

DIN An abbreviation meaning Deutsche Industrie Norm. It is the German system for rating the film stock's sensitivity to light or film speed. It may be used in place of ASA or E.I.

Diopter A filter that allows you to focus on something much closer than the lens would normally allow without the diopter. They are available in various strengths, and the higher the number of the diopter, the closer you can focus. A diopter requires no exposure compensation.

Director of Photography (D.P.) The person in charge of lighting the set and photographing a film. He or she works closely with the Director to transform the written words of the script to the screen, based on the Director's vision. The D.P. oversees all aspects of the camera department and the camera crew. The D.P supervises all technical crews on a production during filming. The D.P. is also called the *cinematographer* or *cameraman.*

Displacement Magazine A magazine that contains the feed and take-up sides on the same side of the magazine. When the magazine is placed on the camera, the feed side is toward the front and the take-up side is toward the rear. A displacement magazine may be of the single-chamber type, which contains both the feed and take-up in the same compartment, or a double-chamber displacement magazine, which has separate compartments for the feed and take-up sides.

Ditty Bag A canvas bag containing many different-sized compartments, which is used by the Camera Assistant to hold tools and supplies needed for filming. Some of the items kept in the ditty bag include basic tools, the slate, tape measure, pens, sharpies, camera tape, etc.

Dolly A four-wheeled platform onto which the camera is mounted for moving shots. It may also have a boom arm that allows the camera to be raised or lowered for a shot.

Donut A circular piece of rubber of various sizes, approximately 1/2-inch thick, with a circle cut out of the center. It is placed on the front of the lens and is used to seal the opening between the lens and the matte box. The donut prevents any light from entering the matte box from behind the lens, and reflecting off the filters and into the lens.

Dummy Load A short roll of raw stock film that is too small to be used for photographing any shots. It may be used to test the magazines for scratches during the camera prep.

Dutch Angle Framing a shot so that the camera is tilted either left or right so that the image will appear diagonally within the frame.

Dutch Head A special type of head that is usually attached to the standard fluid head and allows you to shoot Dutch angle shots. It is set up so that the tilt action is opposite from the tilt of the regular head, and so gives the effect of the Dutch angle.

Eastman Kodak Trade name of a brand of professional motion picture film stock. Also referred to as Kodak.
80A Filter Conversion filter that is used to convert daylight-balanced film for filming with tungsten light sources. When using this filter you must adjust your exposure by 2 stops. It is blue in color.
85 Filter Conversion filter used to convert tungsten-balanced film for filming under daylight conditions. When using this filter you must adjust your exposure by 2/3 stop. It is orange in color.
E.I. Abbreviation for Exposure Index. See Exposure Index.
Emulsion The part of the film stock that is sensitive to light. The emulsion is where the photographic image is recorded.
End Slate See Tail Slate.
Enhancing Filter A filter used to improve the color saturation of red and orange objects in the scene while having very little effect on other colors. It requires an exposure compensation of 1 1/2 to 2 stops.
Expendables Items such as tape, pens, markers, batteries, etc. that are used by the camera department in the daily performance of the job. They are called *expendables* because they are usually used up during the course of a production.
Exposed Film Any film that has been run through the camera and contains a photographed image. Exposed film must be kept in a cool, dark place and opened only in a darkroom.
Exposure The f-stop that has been set for a particular shot. It can also be used to refer to the act of subjecting the film to light. The degree of exposure is determined by how much light strikes the film and for what length of time the light is allowed to strike the film.
Exposure Index (E.I.) A numeric value assigned to a film stock that is a measurement of the film's speed or sensitivity to light. The higher the number, the faster the film and the more sensitive it is to light. It may be used in the same way as the ASA number or DIN number.
Exposure Meter A measuring device used to determine the amount or intensity of light falling on a scene. The two main types of exposure meters are incident and reflected. A spot meter is the most common type of reflected light meter. See Incident Meter, Light Meter, and Spot Meter.
Exposure Time The amount of time that each frame of film is exposed to light. For normal motion picture photography, the standard exposure time is 1/50th of a second with a film speed of 24 frames per second and a 180° shutter angle.
Eyebrow A small flag that mounts directly to the matte box and is used to block any light from hitting the lens. It may also be called a *sunshade* or *French flag*.

Eyepiece The attachment on the camera that allows the Camera Operator to view the scene as it is being filmed. On most modern film cameras, a mirror shutter directs the image entering the lens to the eyepiece for the operator to view. It contains an adjustable diopter to compensate for the differences in each person's vision, and some type of rubber eyecup for comfort and to protect the operator's eye.
Eyepiece Covers A small round cover, with a hole in the center, usually made of foam or chamois material, and placed on the eyepiece so that it is more comfortable for the Camera Operator.
Eyepiece Extension A long version of the camera eyepiece used when a short eyepiece is not convenient or comfortable for the Camera Operator. It is used most often when the camera is mounted to a geared head.
Eyepiece Heater A heater element used to keep the eyepiece warm when shooting in cold weather situations. It prevents the eyepiece from fogging. It may also be called an *eyepiece warmer.*
Eyepiece Leveler A long, adjustable rod that is used to keep the eyepiece level while the camera is tilting. The eyepiece leveler allows the eyepiece to remain at a comfortable position for the Camera Operator when doing extreme tilt moves with the camera.

Feed Side The side of the magazine that contains the fresh, unexposed film.
Field of View The angle covered by the camera lens. It may also be called *angle of view.*
Film Can A metal or plastic container that the fresh raw stock is packaged in. It is also used along with the black bag to wrap any exposed film or short ends created during shooting.
Film Plane The point located behind the lens where the film is held in place during exposure. It is the plane where the rays of light entering the lens come together in sharp focus.
Film Speed The rating assigned to the film based on its sensitivity to light. The film speed is expressed as ASA, DIN, E.I., or ISO.
Film to Video Synchronizer A device used when filming a video monitor image with a film camera. It is needed so that you may eliminate the moving bars on the screen, which are common when filming a video image with a film camera.
Filter A piece of optically correct glass or gel that is placed in front of a lens or light source to cause a change in the image or a change in the light. The filter may be a special color or have a particular texture that gives the desired effect.

Filter Trays Compartments used to hold a filter in the matte box. They usually slide in and out of the matte box and come in various sizes to accommodate different filter sizes.

First Camera Assistant, First Assistant Cameraman (1st A.C.) A member of the camera crew whose duties include overseeing all aspects of the camera department during filming, setting up and maintaining the camera, changing lenses and filters, loading film into the camera, keeping the camera in working order, and maintaining focus during shooting. The 1st A.C. works very closely with the Director of Photography and the Camera Operator during filming.

Flare A bright spot or flash of light in the photographic image that may be caused by lights shining directly into the lens or by reflections from shiny surfaces.

FLB Filter Filter that is used when shooting under fluorescent lights with indoor type B films. It requires an exposure compensation of one stop.

FLD Filter Filter that is used when filming under fluorescent lights with daylight type film. It requires an exposure compensation of one stop.

Fluid Head A mounting platform for the camera that allows the Camera Operator to do smooth pan and tilt moves during shooting. Its internal elements contain a highly viscous fluid that controls the amount of tension on the pan and tilt parts of the head. It may be attached to a high hat, tripod or dolly and is usually operated by using a handle that is mounted to the side of the head.

Foam-tip Swab Similar to cotton swabs. It is a long wooden or plastic stick that contains a small foam swab on one end. It may be used to remove any excess oil when oiling the camera.

Focal Length The distance between the optical center of the lens to the film plane when the lens is focused at infinity. Lenses are referred to by their focal length, which is usually expressed in millimeters, such as 25mm, 32mm, 50mm, etc. A specific focal length lens will determine the image size based on the distance from the camera to the subject.

Focal Plane The specific point behind the lens where the image is focused onto the piece of film. As the film travels through the camera, it is held between the pressure plate and the aperture plate in the film gate. This area where it is held while the image is being recorded is the focal plane. Also called *film plane.*

Focal Plane Shutter A rotating shutter located at the focal plane that alternately blocks light from striking the film and then allows light to strike the film. It works along with the mirror shutter of the camera.

When the shutter is open, light strikes the film, forming an image. When the shutter is closed, no light strikes the film.

Focus The point in the scene that appears sharp and clear when viewed through the camera eyepiece. It may also refer to the act of adjusting the lens to produce a sharp image. Any image that is sharp and well defined is called *in focus* and an image that appears soft or fuzzy is called *out of focus*. Focus may be determined by looking through the eyepiece and turning the focus barrel of the lens until the image appears sharp. It also may be determined by measuring the distance from the film plane to the object being photographed.

Focus Chart A special chart that is used when testing photographic lenses. It is used to determine if the lens focus is accurate according to the markings on the side of the lens.

Focus Extension Accessory that attaches to the right side of the follow focus so you can pull focus from either side of the camera. A flexible focus accessory, called a *focus whip* or *whip*, may also be attached to either side of the follow focus mechanism.

Focus Puller A member of the camera crew who is responsible for maintaining focus during a shot. During the rehearsal, measurements are taken to various points on the set. During shooting she adjusts the focus barrel of the lens to correspond to these measurements, so that the image will remain in focus throughout the shot. In the United States, the Focus Puller is usually the same as the First Camera Assistant.

Focus Whip A flexible extension, usually 6 or 12 inches in length, which allows the assistant to step back from the camera and still be able to follow focus for a shot.

Fog Filter Filter that simulates the effect of natural fog. A fog filter will cause any light in the shot to have a flare. No exposure compensation is required.

Follow Focus, Following Focus The act of turning the focus barrel of the lens during the shot, so that the actors stay in focus as they move through the scene. It may also be referred to as *pulling focus*.

Follow Focus Mechanism A geared attachment that mounts to the camera and contains gears that are engaged to teeth on the lens. It enables the First Assistant Cameraman to follow focus or pull focus during the shot.

Footage Counter A digital or dial type of gauge on the camera that shows the amount of film that has been run through the camera.

Format A term that may refer to the shape of the photographed image when it is projected onto the screen or television. It is the ratio of the width to the height and may also be referred to as the *aspect ratio*. See Aspect Ratio. It may also be used to refer to what film gauge you are shooting: 16mm, 35mm, or 65mm.

Four Inch Lens (4-inch Lens) A slang term used in the early days of filmmaking to indicate a 100mm lens. The term is still used today by some cameramen.

f.p.s. Abbreviation for Frames per Second. See Frames per Second.

Frames per Second The standard measurement for film speed as it runs through the camera or projector. The standard film speed rate when filming with synchronous sound recording is 24 f.p.s.

Frame Rate The speed that the film runs through the camera. It is expressed in terms of Frames per Second (f.p.s.).

French Flag A small flag that is mounted onto the camera and used to help keep any lights from causing a flare in the lens. It consists of a flexible arm, onto which the small flag is attached, and is positioned so that the flag keeps the flare from striking the front element of the lens.

Front Box A wooden box that attaches to the front of the camera head and is used to hold a variety of tools and accessories. It is mainly used by the First Assistant Cameraman for storing the tape measure, mag light, depth-of-field charts, pens, sharpies, compressed air, gum, mints, etc.

F-Stop The setting on the lens that indicates the size of the aperture. It is an indication of the amount of light entering the lens and does not take into account any light loss due to absorption. The f-stop of a lens is determined by dividing the focal length of the lens by the diameter of the aperture opening. The smaller the number, the larger the opening. The standard series of f-stop numbers is as follows: 1, 1.4, 2, 2.8, 4, 5.6, 8, 11, 16, 22, 32,. . . .

Fuji Trade name for a brand of professional motion picture film stock.

Full Aperture The entire area of the film frame that extends out to the perforations on the film. When looking through the eyepiece, it extends beyond the frame lines inscribed on the ground glass.

Gaffer Tape (2″) Cloth tape, usually two inches in width, that may be used for any taping job that requires tape wider than the standard one-inch camera tape. It is available in many colors, including gray, black, and white.

Gate The part of the camera where the film is held while it is being exposed. When speaking of the gate, we are talking about the opening that is cut into the aperture plate, allowing light to pass through so that it strikes the film and creates an exposure on the film. Sometimes when referring to the gate, we include the aperture plate, pressure plate, pull down claw, and registration pin.

Geared Head A mounting platform for the camera that allows the Camera Operator to do smooth pan and tilt moves during shooting. It may be mounted to a high hat, tripod, or dolly, and it is operated by turning two control wheels that are connected to gears in the head. One control wheel is mounted on the left side and is used for panning the camera. The other control wheel is mounted toward the back of the head and is used for making any tilt moves.

Good (G) Any takes that the Director chooses as his or her preference for each scene. These takes are circled on the camera report. It may also refer to the total amount of footage for all takes on the camera report that are circled and are to be printed or transferred by the lab.

Graduated Filter, Grad A filter that is of varying density so that one half of it is clear and the other half contains the filter. Some of the most common graduated filters are the graduated neutral density filters and color graduated filters.

Gray Scale, Gray Card A standard series of tonal shades ranging from white to gray to black. The card or scale may be photographed at the beginning of each film roll and is used by the lab when processing the film to check for the correct tonal values in the film.

Grease Pencils Erasable pencils used for making focus marks directly on the lens or focus marking disk. One popular brand of grease pencils is the Stabilo brand, and they are available in white, yellow, red, and black.

Ground Glass A small piece of optical material, onto which a portion of the light from the lens is focused to allow the Camera Operator to see the image that the lens is seeing. It is usually inscribed with lines to indicate the sides of the frame, which assists the operator in composing the shot.

Guild Kelly Calculator Trade name for a brand of depth-of-field calculator used by many First Camera Assistants.

Hair A very fine piece of emulsion that appears in the gate and can look like an actual hair. If not removed from the gate, it will appear as a large rope on the screen. It may be caused by the emulsion being scraped off of the film as it travels through the gate.

Handheld See Camera, Handheld.

Handheld Accessories Any item needed to make handheld shots easier. These items may include left- and right-hand grips, a shoulder pad, a smaller clamp-on style matte box, and smaller film magazines. The hand grips are used to make it easier to hold the camera, and the shoulder pad makes the camera more comfortable when resting it on your shoulder.

Hard Mattes Covers that are placed in front of the matte box to block any unwanted light from striking the lens. They have the center portion cut out in various sizes depending on the focal length of the lens being used.

Harrison & Harrison Trade name of a brand of motion picture camera filters.

Head A platform for mounting the camera that allows the Camera Operator to make smooth pan and tilt moves during the shot. The two most commonly used heads are *fluid heads* and *geared heads.*

Head Slate A slate that is photographed at the beginning of a shot. When doing a head slate, the slate is held right side up.

High Angle A shot that is done with the camera placed very high above the action and pointed down toward the subject or action. When doing a high angle shot the camera may be placed on the tripod, on a ladder, or even on a crane boom arm.

High-Hat (Hi Hat) A very low camera mount used when filming low angle shots. The head is mounted to the high hat and then the camera is mounted onto the head.

High Speed Any filmed shot that is done at a speed greater than normal sync speed of 24 f.p.s. There are many specialized cameras that allow you to film at high speeds, anywhere from 300 to 1000 or 2000 f.p.s. When filming at high speed, the final projected film image will have the illusion of moving at a much slower rate of speed. It is very useful for slowing down fast action.

HMI Lights Lighting devices that produce a color temperature that is equivalent to the color temperature of natural daylight. They are often used when filming daylight interior scenes, to help supplement the existing daylight coming through the windows.

HMI Speed Control A speed control that may be used when filming with HMI lights. HMI lights are balanced for daylight, and when using them you can only film at certain speeds, otherwise there will be a flicker in the photographic image.

Hyperfocal Distance The closest point in front of the lens that is in acceptable sharp focus when the lens is focused at infinity. This distance is determined by the focal length of the lens and the f-stop set on the lens. If you set the focus of the lens to the hyperfocal distance, everything from one-half the distance to infinity will be in focus. There are tables listing the hyperfocal distance for various lenses and f-stops that are published in many film books.

I.A.T.S.E. Abbreviation for International Alliance of Theatrical and Stage Employees.
Inching Knob A small knob that may be located either inside or outside the camera body that allows you to slowly advance the film through the movement. It is most often used at the time of threading to check that the film is traveling smoothly and not binding or catching anywhere.
Incident Light The light from all sources that falls on the subject being filmed. To measure the amount of incident light falling on a subject, you would use an incident light meter.
Incident Meter A light meter used to measure the amount of incident light that is falling on the subject.
Insert Slate A small scene slate used to identify any MOS or insert shots being filmed. The information written on it includes the production title, roll number, scene and take numbers, date, Director's name, and Director of Photography's name. The insert slate is different from the sync slate because it usually does not contain the clapper sticks.
Intermittent Movement The starting and stopping movement of the film transport mechanism as it advances the film through the camera.
Iris An adjustable diaphragm that is used to control the amount of light that is transmitted through the lens and exposes the film. The iris of the lens consists of overlapping leaves that form a circular opening to vary the amount of light coming through the lens. By changing the f-stop or t-stop setting of the lens, you are changing the iris opening of the lens.
Iris Rods Metal rods of varying lengths that are used to support the matte box, follow focus or other accessory on the camera.
ISO Abbreviation for International Standards Organization. It is a rating of the film stock based on its sensitivity to light. It is sometimes used in place of ASA or E.I.

Kimwipes Soft tissue-like material, similar to lens tissue. They can be used for cleaning filters or any other small cleaning job, but should not be used to clean lenses.
Kodak Trade name for a brand of professional motion picture film stock. Also known by its full name, Eastman Kodak.

Lab The facility where the film is sent to be processed, developed, and printed.

Latitude The ability of the film emulsion to be underexposed or overexposed and to still produce an acceptable image. Many of the newer, faster speed films have a greater latitude than the earlier slow-speed film stocks.

Left-Hand Grip An attachment for the camera used when shooting handheld shots. It is placed on the left side of the camera and allows the Camera Operator to hold the camera steady in a comfortable position for shooting.

Legs A slang term used to refer to the tripod for the camera. *Baby legs* refers to the smaller tripod, and *standard legs* refers to the larger size tripod.

Lens An optical device through which light rays pass to form a focused image on the film. Lenses are commonly referred to by their focal length, which is expressed in millimeters. The two main types of lenses are prime and zoom. One type of prime lens is the telephoto lens. It is a lens of an extremely long focal length, such as 200mm, 300mm, 400mm, and 600mm.

Lens Cleaner Liquid that is used to clean lenses and filters along with lens tissue. Lens cleaner should be used to clean lenses only when absolutely necessary.

Lens Extender An attachment that is mounted between the lens and the camera that increases the focal length of the lens being used. The most common lens extenders are the 1.4x, which increases the focal length by 1.4 times the actual focal length, and the 2x, which doubles the actual focal length. When using a lens extender you must adjust your exposure by the strength of the extender. For example, when using a 2x extender, you must adjust your exposure by 2 stops.

Lens Light A small light, mounted to a flexible arm that is attached to the camera, which allows the First Assistant to see the lens focus and zoom markings when filming in a dark set. It is sometimes called a *little light* or *niner light*.

Lens Shade A rubber or metal device that either screws on or is clamped onto the front of the lens. It is used to hold round filters and to keep any stray light from striking the front element of the lens. It may also be called a *sunshade*.

Lens Speed The lens speed refers to the widest f-stop that the lens opens up to. The smaller the f-number, which means the larger the opening of the lens aperture, the faster the lens. Fast lenses allow you to film in very low light situations.

Lens Tissue Small tissues used to clean lenses and filters along with lens cleaner. It is recommended that you never use a dry piece of lens tissue on a lens because it may scratch the coating or front element of the lens.
L-Handle See Speed Crank.
Light Meter A measuring device that is used to measure the amount of light that is illuminating the scene. See Exposure Meter, Incident Meter, Reflected Meter, and Spot Meter.
LLD Filter A filter used when filming with tungsten-balanced film in low-light daylight situations. It is often used in place of the 85 filter. It requires no exposure compensation.
Loader The member of the camera crew who is responsible for loading and unloading the film into the magazines. A loader is usually used on larger productions when two or more cameras are being used.
Lock Off Any shot that is done with the pan and tilt mechanisms of the camera head locked so that the camera is not moved during filming. It may be used for a stunt or special effects shot.
Long Lens Term used to refer to a telephoto lens or a lens that has a focal length that is longer than that of a normal lens.
Loop A section of film between the gears of the camera or magazine, set to a specific length that allows for the intermittent movement of the film through the camera during filming. If the loop is not set correctly, the film may become jammed in the camera or magazine, and the camera will not run properly or it may run loudly.
Low Contrast Filter A filter that lowers the contrast by causing light to spread from highlight areas to shadow areas. It will mute colors and make blacks appear lighter, and also allows more detail in dense shadow areas. Also referred to as a *lo-con filter.* No exposure compensation is required.
Low Angle A shot that is done with the camera placed very close to the ground and pointed up toward the subject or action. When doing a low angle shot, the camera may be placed directly on the ground, placed on a high hat or on a rocker plate.
Low Hat A very low camera mount used when filming low angle shots. It is similar to the high hat, only it enables you to get the camera lower. The head is mounted to the low hat, and then the camera is mounted onto the head.

Mag Abbreviation for Magazine.
Magazine A removable, lightproof container that contains the film before and after exposure. The two main types of film magazines are

the displacement magazine and the coaxial magazine. There are two distinct areas in each film magazine. The feed side, which contains the fresh unexposed film, and the take-up side, which contains the exposed film. See Coaxial Magazine and Displacement Magazine.

Magliner The trade name of a four-wheel, folding hand truck used by many camera assistants to expedite the moving of the many equipment cases on a film set. The two most common types of magliner carts are the Gemini Junior and the Gemini Senior.

Mag Lite Small pocket-type flashlight used by most crew people. It has a very bright and powerful bulb, and by turning the head of the light you are able to adjust the beam from spotlight to floodlight.

Marks Small pieces of colored tape or chalk marks that are placed on the ground, and are used to identify various positions. They may indicate where the actor is to stand for the shot, where the dolly starts and stops its move, or as a reference for focus used by the First Camera Assistant. The most common type of actor's mark is the T-mark.

Matte Box An accessory that mounts to the front of the camera to shield the lens against unwanted light and also used to hold any filters. It is usually referred to by the size of filters that it normally holds. For example, a matte box that holds 4-inch by 4-inch filters is called a 4 × 4 matte box. In addition to holding two or three square or rectangle filters, most matte boxes have a snap in piece that holds one round filter, either 4 1/2 inches in diameter or 138mm in diameter.

Mirrored Shutter A shutter that incorporates a mirror into its design. When the shutter is open, the light goes to the film and the image is recorded on film. When the shutter is closed, the light strikes the mirror and the image is directed to the viewfinder for the Camera Operator to view the shot.

Mitchell A trade name of one of the earlier models of motion picture cameras. It was mainly used for studio work because of its large size. Some of the models of the Mitchell camera include the BNC and BNCR.

Mitchell Flat Base A type of top casting of the high hat, low hat, tripod, or dolly onto which the head is mounted.

Mitchell Diffusion The trade name of a brand of motion picture camera diffusion filters. As with other diffusion, they are used to soften or decrease the sharpness of the image. Mitchell Diffusion filters are available in various densities from very light to very heavy. They require no exposure compensation. See Diffusion Filters.

Monitor A television or video screen used by the Director during filming to check the framing of the shot and the quality of the performance. It is used in conjunction with a video camera that is attached to the film camera.

MOS Any shot that is done without recording synchronous sound. It is an abbreviation for "Minus Optical Sound."

Multi-camera The term given to any shots that are done by using more than one camera during filming. The different cameras will record the same action of the scene, from different angles.

Mutar A 16mm lens attachment that is designed for use on a Zeiss 10mm to 100mm zoom lens. It is used to increase the angle of view of the lens and to make it more wide angle than it already is. The lens is placed in its macro setting before the mutar is attached. Focus is then adjusted by turning the zoom barrel of the lens instead of the focus barrel.

ND Abbreviation for Neutral Density filter.

Negative Film that when processed produces a negative image of the scene. A print must be made of this negative for viewing purposes. This term is sometimes used to refer to unexposed raw stock used for most film productions.

Neutral Density Filter (ND) A filter used to reduce the amount of light that strikes the film. It has no effect on colors. Neutral density filters are gray in color and come in varying densities. Neutral density filters require an exposure compensation depending on the number and density of the filter. ND3, one stop; ND6, two stops; ND9, three stops.

Nitrate Base Film A highly flammable film stock used in the early days of filmmaking. It was made up of cellulose nitrate, which was capable of self-igniting under certain circumstances. It is no longer used for the manufacture of motion picture film.

No Good (NG) Any take that is not printed or circled on the camera report. There may have been a technical problem during shooting or possibly the performance was not acceptable to the Director. On the daily film inventory report form it refers to the total amount of footage for all takes on the camera report that are not to be printed or transferred by the lab.

Obie light A light that is mounted on the camera directly over the matte box. It is used to highlight the actor's eyes.

O'Connor Trade name for a brand of professional motion picture fluid head.

1.85 (One-Eight-Five) The standard aspect ratio for today's theatrical motion pictures. It may also be written as 1.85:1, which means that the picture area is 1.85 times as wide as it is high. When filming in 1.85 format, standard spherical lenses and equipment may be used.

One-Inch Lens (1-Inch Lens) A slang term used in the early days of filmmaking to indicate a 25mm lens. The term is still used today by some cameramen.
Opening Up the Lens Turning the diaphragm adjustment ring on the lens to a smaller f-stop number that results in a larger diaphragm opening. Opening up allows more light to strike the film.
Operator See Camera Operator.
Optical Flat A clear piece of optically corrected glass that is placed in front of the lens to protect the lens. The optical flat may also be used to help reduce the sound coming from the camera. It is available in the same sizes as most filters and should be included in the filter order on any film production.
Orangewood Sticks Wooden sticks that are used to remove emulsion buildup in the gate or aperture plate. The aperture plate or gate should only be cleaned with these sticks. It is advisable to never use any type of metal or sharp instrument when cleaning the gate.
Overcrank Running the camera at a speed that is higher than normal sync speed. This causes the action to appear in slow motion when it is projected at sync speed of 24 f.p.s. The term was originated in the early days of filmmaking, when all cameras were cranked by hand.
Overexpose Allowing too much light to strike the film as it is being exposed. This results in the photographic image having a washed-out look or being much lighter than normal. The Director of Photography may overexpose the film in order to achieve a desired effect. It is much better to overexpose the film stock than to underexpose it, because you will have more latitude to correct any problems when printing or transferring the final film.

Pan or Panning The horizontal or left and right movement of the camera. The camera is usually mounted to either a geared head or fluid head, which is on the tripod or dolly. By panning the camera during the shot, the Camera Operator is able to follow the action within the scene.
Panavision Trade name of a brand of professional 16mm and 35mm camera.
Paper Tape (1/8″ or 1/4″) Tape that is most often used to make focus marks on the lens. It is wrapped around the barrel of the lens so that you may mark it for following focus.
Paper Tape (1/2″ or 1″) Tape that may be used for making actor's marks, labeling equipment, or any other taping job during production.
Paper Tape (2″) Tape that is used for the same types of things as gaffer tape. It may be used to seal any cracks or holes in the darkroom.

Perforations, Perfs Equally spaced holes, punched into the edges of the film along the entire length of the roll. These holes are engaged by the teeth of the sprockets, in the film magazines and camera movements, allowing the film to accurately travel through the camera before and after exposure. In 35mm film there are four perforations per film frame, on each side, and in 16mm there are two per frame on each side.
Persistence of Vision The phenomenon that allows the human eye to retain an image for a brief moment after it has been viewed. This allows the illusion of movement when a series of still pictures are projected on a screen at a specified rate of speed. At normal sync speed of 24 f.p.s., a series of still frames projected on the screen appear to be moving continuously to the human eye.
Pitch The distance between the bottom edge of one perforation to the bottom edge of the next perforation. This distance is measured along the length of the film.
Polarizing Filter A filter that is used to reduce glare and reflections from reflective, nonmetallic surfaces. It may be used when filming into or through a window. It is also used to enhance or darken a blue sky or water. It is mounted so that it can be rotated until the desired effect is achieved. While looking through the lens, the Director of Photography or Camera Operator rotates the filter until the correct or desired effect is achieved. An exposure compensation of 1 1/2 to 2 stops is required.
Powder Puffs Soft makeup-type pads that are used to erase information placed on an acrylic slate with erasable slate markers.
Precision Speed Control This is a speed control attachment that allows you to vary the speed of the camera. It enables you to vary the speed to a precise degree, sometimes to three decimal places. It may be used when filming high speed or slow motion.
Prep The time during pre-production when the equipment is checked to be sure that it is in working order. During the camera prep, the First Camera Assistant goes to the camera rental house and sets up and tests all of the camera equipment to be sure that it is in proper working order for filming.
Pressure Plate A flat, smoothly polished piece of metal that puts pressure on the film, keeping it flat against the aperture plate and steady as it travels through the gate. It is usually spring loaded and is most often located behind the aperture plate inside the camera, but it may also be a part of the magazine.
Primary Colors For the purposes of cinematography, the three primary colors of light are red, blue, and green. When equal amounts of

these three colors of light are combined, they form what is known as white light. All colors of light are made up of varying combinations of these primary colors. The corresponding complementary colors to these are cyan, yellow, and magenta, respectively.

Prime Lens A lens of a single, fixed focal length that cannot be changed. Examples of prime lenses are 25mm, 35mm, 50mm, 75mm, 100mm, etc.

Print All The instructions given to the lab that tells them to print all of the takes on a given roll of exposed film.

Print Circle Takes Only The instructions given to the lab that tells them to print only the takes that have been circled on the camera report, for a specific roll of film.

Production Company The name of the company that is producing the film.

Production Number The number of the film or television episode as assigned by the production company.

Production Title The working title of the film as assigned by the production company.

Professional Cameraman's Handbook An indispensable manual used by both Camera Assistants and Director's of Photography. It contains illustrations and descriptions of the many different cameras and related pieces of equipment in use today. For each camera, it describes how to load the magazine with film and how to thread it in the camera.

Pro Mist Filter A diffusion filter that is used to soften harsh lines in an actor's face. It may sometimes give the illusion of the image being out of focus. Pro Mist filters come in varying densities and are available in white and black diffusion. No exposure compensation is required.

Pull Down Claws These are the small hooks or pins, located in the camera movement, which reach into the perforations of the film and pull the film down into position in the gate so that it may be photographed.

Pulling Focus See Following Focus.

Raincover A waterproof cover used to protect the camera in extreme weather conditions, including snow and rain. It has an opening for the viewfinder eyepiece and lens, and contains clear panels for the assistant to view the lens markings.

Raw Stock Fresh, unexposed and unprocessed film stock.

Reflected Light Any light that is bouncing off, or being reflected by an object. Reflected light is usually measured with a reflected light meter or spot meter.

Reflex Viewing System A viewing system that allows the Camera Operator to view the image as it is being filmed. The image is reflected to the viewfinder, usually by a mirror shutter, allowing the Camera Operator to see the image through the lens. Because the image is sent to the viewfinder intermittently, there is a slight flicker in the viewfinder while the camera is running.

Registration The accurate positioning of the film in the film gate as it is running through the camera. Any variation will cause a jump or blur in the photographic image. During the camera prep, the registration may be checked by filming a registration chart and then viewing the results.

Registration Chart A chart containing a series of crossed lines that is used during the camera prep to check the registration of the camera. By shooting a double-exposed image of the chart and then viewing the results, you can tell if the registration of the camera is accurate.

Registration Pins Part of the camera movement consisting of a small metal spike or pin that holds the film securely in the gate while it is being exposed. Most cameras contain a single registration pin, while some of the cameras used for special effects cinematography contain two registration pins.

Remote Switch A camera on/off switch that is usually connected to a long cable running from the camera allowing the starting and stopping of the camera from a distance.

Reversal Film that when processed produces a positive image of the scene. It may also be called *positive film*, and it may be viewed directly.

Right-Hand Grip A camera accessory item used when filming handheld shots. As the name implies, it attaches to the right-hand side of the camera and is used to hold the camera steady during shooting. Most right-hand grips contain an on/off switch for the Camera Operator to start and stop the camera.

Rocker Plate A low-angle camera mount, usually consisting of two separate pieces. The camera mounts to the top section that has a curved base, allowing it to rock back and forth. This top section fits into the bottom cradle that swivels left and right. The combined movement of rocking the camera back and forth, and swiveling the base left and right, allows the Camera Operator to make smooth pans and tilts from very low angles.

Roll Number The number assigned to a roll of film when it is placed on the camera. Each time a new roll of film is placed on the camera, the next higher number is assigned to that roll.

Ronford Trade name of a brand of professional motion picture fluid heads and tripods.
Rubber Donut See Donut.
Rushes See Dailies.

Sachtler The trade name of a brand of professional motion picture camera fluid heads and tripods.
Samuelson MKII Calculator Trade name of a brand of depth-of-field calculator used by most First Camera Assistants.
Scene Number The number assigned to a scene based on its place in the script. A scene is a section of the film as it takes place in a particular location or time in the story. Normally each time the location or the time changes, a new scene number is assigned to the action. For slating purposes, the scene number is written on the slate and also on the camera report as a reference to which scenes have been filmed on which roll of film.
Script Supervisor The person on the film crew who keeps track of the action for each scene. He or she keeps notes for each shot regarding actor movement and the placement of props. The script supervisor tells the Second Camera Assistant what the scene and take numbers are for each shot.
Second Camera Assistant, Second Assistant Cameraman (2nd A.C.) The member of the camera crew whose duties include assisting the First Camera Assistant, clapping the slate for the shot, keeping camera reports, placing marks for actors, and loading and unloading film into the magazines. The 2nd A.C. reports directly to the 1st A.C. during production.
"Second Marker" What the 2nd A.C. calls out when slating a shot if the first slate was missed by the Camera Operator or Sound Mixer.
Second Camera An additional camera used for filming shots or scenes at the same time as the primary or main camera. Many productions are filmed using only one camera, moving it for each new shot or setup. By using a second camera, you are able to film two different shots at the same time, for example, a master shot and a close-up. The second camera may also be used when filming shots that can only be staged once, such as car stunts, fire shots, explosions, etc.
Shoulder Pad A small pad that attaches to the underside of the camera when doing handheld shots. It is used to make the Camera Operator more comfortable by allowing him to hold the camera on his shoulder for longer periods of time without becoming fatigued.

Short End (S.E.) A roll of unexposed raw stock that is less than a full size roll but larger than a waste roll or dummy load. In 16mm, a short end is usually any roll larger than 40 feet and in 35mm it is usually any roll larger than 100 feet.
Short Eyepiece A smaller version of the camera eyepiece that is used especially when filming handheld shots. It may also be used on the camera in certain filming situations where the long eyepiece is too uncomfortable or in an awkward position.
Shutter The mechanical device in a camera that rotates during filming to alternately block light from the film and then allow it to strike the film. Many shutters contain a mirror that reflects some of the light so that it enters the eyepiece, allowing the Camera Operator to view the shot during filming. As the shutter turns, it alternately blocks the light from the eyepiece, allowing it to strike the film, and then blocks the light from hitting the film, allowing it to go to the eyepiece. This causes a slight flicker in the eyepiece when viewing the shot during filming.
Shutter Angle A measurement in degrees of the opening of the camera shutter. For most professional cinematography, the standard shutter angle is 180°. Many cameras contain fixed shutters that cannot be changed or adjusted. Some cameras have an adjustable shutter angle to allow for special types of filming.
Shutter Speed The exposure time of the shutter while the camera runs at a specific speed. The standard shutter speed or exposure time for motion pictures is 1/50th of a second at a sync speed of 24 f.p.s., with a shutter angle of 180°.
Silicone A type of lubricant that is available in a spray or liquid form. The spray is used to lubricate the sliding base plate or tripod legs if they begin to stick. It should never be used on any of the moving parts of the camera motor or movement. The liquid type is usually used to lubricate the pull down claws of some cameras, including the Panavision and Ultracam.
16mm A film gauge, introduced in 1923, that was used mainly for nontheatrical or amateur productions. It is most commonly used today for music videos, commercials, and many television series. Because of the lower price of the film stock, as well as the lower price of renting the camera equipment, it provides a less expensive filming alternative to using 35mm film. For television broadcast or videocassette release, it can provide very good results. 16mm film contains 40 frames per foot and contains two perforations per frame on each side of the film. At normal sync speed of 24 f.p.s., it will travel through the camera at the rate of 36 feet per minute.

65mm/70mm Film gauge that is most often used for release prints of theatrical films. It is very rarely used for actual productions. There are some 65mm cameras available for filming for any production company that would like to shoot with 65mm. Arriflex and Panavision are two companies that still manufacture 65mm cameras for filming.

Slate A board marked with the pertinent identifying information for each scene photographed. It should contain the film's title, Director's name, Cameraman's name, date, camera roll number, scene number, and take number. The two main types of slates are sync and insert. See Insert Slate and Sync Slate.

Slate Markers Erasable markers that are used to mark information on acrylic slates. They are usually some type of dry erase marker. The two most common brands of slate markers are EXPO and WRITE ON/WIPE OFF markers.

Sliding Base Plate An attachment used for mounting the camera to the head. It is usually a two part plate with the bottom piece mounted to the tripod head, and the top piece mounted to the camera. It allows you to slide the camera forward or backward on the head to achieve proper balance. By achieving proper balance, the Camera Operator does not have to struggle in order to make smooth pan and tilt moves.

Soft Contrast Filter A filter that lowers the contrast by darkening the highlight areas. Also referred to as a *soft-con filter*. No exposure compensation is required.

Soft Focus Indicates a shot or scene that appears to be out of focus to the viewer's eye. It may be caused by the First Camera Assistant or Focus Puller not setting the lens focus properly, so that the image is fuzzy and lacks any sharp detail around the edges. It may also refer to a shot that contains some type of a diffusion filter in front of the lens in order to create a specific soft or dreamlike effect.

Space Blanket A large cover used to protect the camera and equipment from the sun and weather. It is usually a bright silver color on one side and may be red, green, blue, or another color on the opposite side.

Speed Crank An "L"-shaped handle that attaches to the follow focus mechanism. It is used when the First Camera Assistant or Focus Puller has a very long focus change to do during a shot.

Speed (Camera) The rate at which the film travels through the camera. Standard film speed in the United States is 24 frames per second (f.p.s.). For 16mm film this translates to 36 feet per minute and for 35mm film it is 90 feet per minute.

Speed (Film) This is an indication of the film's sensitivity to light. The film speed may be referred to by the ASA, DIN, or E.I. number. A fast film has a higher number, while a slow film has a smaller number. The higher the number, the more sensitive the film is to light.

Speed (Lens) The f-stop or t-stop setting of the lens at its widest opening. The smaller this number is, the faster the lens is. Fast lenses are used many times for filming in extreme low light situations.

Split Diopter A type of filter that may be used to maintain focus on two objects, one in the foreground and one in the background. The split diopter is round in shape, and only half of it contains the diopter. The remaining half of the filter is clear. No exposure compensation is required.

Split Focus The technique of setting the focus so that a foreground object and a background object are both in focus for a shot or scene. By setting the focus to a point in between the two objects, you may be able to keep them both in focus. Split focus is not possible in every shooting situation. It will depend on the distance of the subjects, the lens used, the f-stop set on the lens, and the depth of field for the particular situation. It is usually a good idea to check with the Director of Photography to see if he or she wants you to try to do a split focus for a shot.

Spot Meter An exposure meter that takes a light reading by measuring the light that is reflected by an object. It is called a *spot meter* because it takes its measurement from a very small area of the object. This small area is called the *angle of acceptance*, and in most cases it is around 1°. When using a spot meter, it is standard practice to stand at the position of the camera and point the meter toward the object.

Spreader A metal or rubber device that has three arms and opens up to form a horizontal Y shape to support the legs of the tripod. It prevents the legs of the tripod from slipping out from under the camera. It may also be called a *spider* or a *triangle*.

Sprockets Small teeth or gears inside the camera or projector that advances the film. The wheels that contain the sprockets turn and engage with the perforations of the film, moving it through the camera.

Sprocket Holes Equally spaced holes punched into the edges of film stock so that it may be advanced through the camera or projector. See Perforations.

Stabilo Grease Pencil The brand name of a grease pencil used by many First Camera Assistants. They are available in red, black, white, and yellow.

Standard Legs, Standards The slang term to indicate the tripod on which the camera and head are mounted. They are called *standards* because they are the tripod that are most commonly used when not mounting the camera on a dolly. Most standard tripods can be adjusted in height from approximately 3 feet to 5 1/2 feet.

Star Filter A filter placed in front of the lens to give highlights to any lights that appear in the scene. The star filter produces lines coming from bright lights in the scene, depending on the texture of the star filter. A four-point star filter will produce a series of four lines coming from a light source, a six-point will produce six lines, and so on. Star filters have a tendency to soften the shot, similar to a diffusion filter. When using any star filter, no exposure compensation is required.

Stick-On Letters (1/2″ or 3/4″) Plastic or vinyl, adhesive-backed letters and numbers that are used to label the slate with production title, Director's name, Cameraman's name, and date. They are available in many colors, including red, blue, green, black, white, and yellow.

Sticks Slang term used to refer to the tripod. Also slang for the sync slate or clap sticks.

Stop An abbreviation meaning the f-stop or t-stop.

Stop Pull The technique of changing the f-stop or t-stop setting of the lens during a shot. The actor may be moving from a brightly lit area to a shady area, requiring a change in the f-stop because of the change in the amount of light that is illuminating the actor.

Stopping Down the Lens See Closing Down the Lens.

Sunshade A small flag or hood that attaches directly to the matte box or the lens to help eliminate any light from striking the lens or the filter. It may also be called an *eyebrow, French flag*, or *lens shade*. See Eyebrow, French Flag and Lens Shade.

Supafrost Filter Trade name of a brand of motion picture camera diffusion filter. It is a type of diffusion filter used to soften the overall image. Great care must be taken when using these filters because they are constructed from plastic and scratch very easily. As with other diffusion filters, they come in varying densities and require no exposure compensation.

Sync Abbreviation for synchronization or synchronized. It is usually used to indicate a film or scene that is shot with sound being recorded simultaneously.

Sync Slate Slate used for identifying all shots done with sound. It contains two hinged pieces of wood that are clapped together at the beginning of each sound take.

Sync Speed The speed that gives motion pictures the appearance of normal motion to the viewer. In the United States, sync speed is 24 frames per second (f.p.s.).

Tachometer A dial or meter located on the camera that shows the speed while the camera is running. The First Camera Assistant watches the tachometer during filming to be sure that it maintains the correct speed.

Tail Slate A slate that is photographed at the end of a shot. When doing a tail slate, the slate is held upside down.

Take, Take Number The number assigned to a scene each time it is photographed. It refers to a single, uninterrupted shot filmed by the camera. Each time a scene or portion of a scene is shot, it is given a new take number.

Take-Up Side The side of the magazine that contains the exposed film. The unexposed film travels from the feed side of the magazine, through the camera, and then into the take-up side of the magazine.

Tape Measure A device used by the First Camera Assistant to measure the distance from the film plane of the camera to the subject. This measurement allows the First Assistant to maintain the focus during filming. The typical tape measure is 50 feet in length and is made of cloth or fiberglass material. It should not be made of metal for safety reasons.

Telephoto Lens A lens of long focal length that allows you to photograph close shots of far away objects. It has a small angle of view. A telephoto lens has a tendency to flatten images and reduce perspective. It can make objects appear closer together than they actually are. Also, when using telephoto lenses, depth of field is greatly reduced.

35mm The standard film gauge, introduced in 1889, that is used for most professional theatrical and television productions. It is used primarily for larger productions because of its excellent image quality. It may also be used for commercials, music videos, and smaller productions in order to have the highest quality image. 35mm film stock contains 16 frames per foot and 4 perforations per frame on each side of the film. At normal sync speed of 24 f.p.s., it travels through the camera at the rate of 90 feet per minute.

Three-Inch Lens (3-inch Lens) A slang term used in the early days of filmmaking to indicate a 75mm lens. The term is still used today by some cameramen.

Tiffen Trade name of a brand of motion picture camera filter.

Tilt The vertical or up-and-down movement of the camera. By turning the tilt wheel of the geared head or moving the handle of the fluid head up or down, the Camera Operator is able to follow any vertical movement of the action in the scene.

Tilt Plate An accessory that is attached between the camera and the head, and is used when doing extreme tilt angles with the camera. It allows the Camera Operator to tilt the camera at a much steeper angle than is possible with the standard geared head or fluid head. Many geared heads contain a built-in tilt plate for these types of shots.

Total (T) A section on the camera report, and also the film inventory form, that indicates the combined total of all Good, No Good, and Waste footage.

Tracking (Lens) The ability of a zoom lens to stay centered on a particular point throughout the range of its zoom.

Triangle See Spreader.

Tripod A three-legged camera support that can be adjusted in height. It is usually made of wood or metal. When choosing a tripod, be sure that its top casting piece is the same as the head that will be used for filming. For example, a tripod with a flat base will not accept a head with a bowl base without some type of adapter piece. See Baby Legs and Standard Legs.

T-Stop A number that is similar to the f-stop, but it is much more precise. It indicates the exact amount of light transmitted through the lens. The f-stop indicates the size of the iris diaphragm opening and does not take into account light loss due to absorption. The standard series of T-stop numbers is 1, 1.4, 2, 2.8, 4, 5.6, 8, 11, 16, 22, 32, 45, 64, etc.

Tungsten Any light source with a color temperature of approximately 3200 degrees Kelvin.

Two-Inch Lens (2-inch Lens) A slang term used in the early days of filmmaking to indicate a 50mm lens. The term is still used today by some cameramen.

Ultracam Trade name for a brand of professional 35mm camera.

Undercrank To operate the camera at any speed that is slower than normal sync sound speed of 24 f.p.s. When the final film is projected, it gives the illusion of faster speed on the screen. It is often used in comedy films for a specific effect. As with the term overcrank, it originated in the early days of filmmaking when all cameras were cranked by hand.

Underexpose Exposing the film to less light than you would for a normal exposure. By allowing too little light to expose the shot, you end up with a very dark image. In general it is not a good idea to underexpose any shot because it is much harder to correct it in the lab during processing.

Variable Shutter A camera shutter that allows you to change the angle for specific filming situations. It allows you to make longer or shorter exposures while the speed of the camera remains constant. On some cameras containing variable shutters, you are able to adjust the shutter angle while the camera is running.

Variable Speed Camera Motor A motor that allows you to change the speed of the camera for certain types of shots. It enables you to film at very slow speeds or very fast speeds, depending on the effect that you want.

Video Assist A system that incorporates a video camera onto the film camera. The image striking the mirror shutter of the camera is split so that part of it goes to the viewfinder and part goes to the video camera. This allows the Director to see the shot on a video monitor as it is being filmed. The shot may also be recorded by connecting a video recorder to the video assist.

Video Cables Any cables needed to connect the video tap to the video monitor or recorder. They may connect the video tap to the monitor or the monitor to the video recorder. They are also called *coaxial cables* or *bnc cables.*

Video Monitor A television monitor that is used along with the video tap to allow the Director to view the shot during filming. See Video Assist.

Video Tap This is a video camera that is attached to the film camera during shooting. It is used by the Director to view the shot on the video monitor as it is being filmed. See Video Assist.

Viewfinder The attachment on the camera that allows the Camera Operator to view the action. Today's modern film cameras all contain what is called a *reflex viewfinder system.* This allows the Camera Operator to line up the shot and view it during filming exactly as it will appear on film. The image coming through the lens is reflected onto a mirror shutter and is formed on a ground glass that is seen through the viewfinder by the Camera Operator. The ground glass allows the operator to focus the image properly. See Eyepiece.

Vignetting Term used to indicate that a portion of the matte box or lens shade is visible or blocking the frame when viewing through the lens. It usually occurs on a very wide angle lens. When looking through the viewfinder, the operator should look around all of the edges of the frame to be sure that nothing is creeping in or blocking any of the frame. In most cases, all that is needed to correct this problem is to slide the matte box back or remove the lens shade, so that it no longer is seen through the viewfinder.

Vinten Trade name of a brand of professional fluid camera head.

Waste (W) The amount of footage remaining on a roll that is left over after the Good and No Good footage has been totaled. It is too small to be called a short end and may be used as a dummy load. In 16mm, waste is usually any roll that is less than 40 feet, and in 35mm it is any roll that is less than 100 feet.

Weaver Steadman The trade name of a brand of professional motion picture fluid head.

Whip A slang term used for a type of follow focus extension. It is called a *whip* because it usually consists of a small round knob attached to a long flexible type cable, which connects to the follow focus mechanism on the camera. It may also be called a *focus whip*.

Wide-Angle Lens A lens that has a very short focal length or a focal length less than that of a normal lens. It may exaggerate perspective and covers a large angle of view. A wide-angle lens is most often used for establishing or long shots, but may be used for a close-up if the intent is to have a distorted image. When using a wide-angle lens, depth of field is greatly increased.

Wrap The period at the end of a day's shooting, or at the completion of the film or production, when all of the equipment is packed away. At the conclusion of filming entirely, the wrap usually consists of cleaning and packing the equipment and returning it to the rental house.

Wratten Filter An optically correct gel filter that is used on a camera lens in place of or in addition to a glass filter. In most cases the gel filter is placed behind the lens in a special gel filter holder. These gels are different than the gels used on lights because they are manufactured to very high standards for use on photographic lenses. They are manufactured by Eastman Kodak and the standard size of Wratten gels is 3 inches × 3 inches square.

Zeiss Trade name of a brand of professional motion picture camera lens.

Zoom The effect achieved by turning the barrel of the zoom lens, to change the focal length of the lens, so that the object in the frame appears to move either closer or farther away from the camera.

Zoom In The act of changing the focal length of the lens so that the angle of view decreases and the focal length of the lens increases. By doing this, the subject becomes larger in the frame. Zooming in decreases depth of field.

Zoom Lens A lens that has varying focal lengths. It allows you to change the focal length by turning an adjustment ring on the barrel of

the lens. Examples are 10mm to 100mm, 12mm to 120mm, 25mm to 250mm, and so on.
Zoom Motor A motor that attaches to the zoom lens to allow you to do a smooth zoom move during a shot. It may be built into the lens, or it may be an additional item that you must attach to the lens.
Zoom Out The act of changing the focal length of the lens so that the angle of view increases and the focal length of the lens decreases. By doing this, the subject becomes smaller in the frame. Zooming out increases depth of field.

Index